TO
MICROPROCESSOR
THEORY & OPERATION

A Self-Study Guide With Experiments

INTRODUCTION
TO
MICROPROCESSOR
THEORY & OPERATION

A Self-Study Guide With Experiments

by
J.A. Sam Wilson
and
Joseph Risse

PROMPT.
PUBLICATIONS

An Imprint of
Howard W. Sams & Company
Indianapolis, Indiana

PROMPT® Publications is an imprint of Howard W. Sams & Company, 2647 Waterfront Parkway, East Drive, Suite 300, Indianapolis, IN 46214-2041

This book was originally developed and published as *Microprocessor Theory and Operation* by TAB Books, Inc., Blue Ridge Summit, PA 17214.

International Standard Book Number: 0-7906-1064-7

Cover Design by: Christy Pierce
Additional Edits by: Tim Clensey

Printed in the United States of America

9 8 7 6 5 4 3

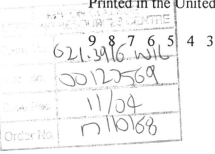

Contents

Introduction

This book provides a smooth transition from digital circuitry to microprocessor systems by using a combination of theory discussions and relatively simple experiments. However, the experiments are optional sections at the end of the chapters. Therefore, the book will also be beneficial to anyone who wants to study the microprocessor system, bit-by-bit, without engaging in the hands-on experiences.

In preparing this book the authors have presumed that the reader is familiar with basic logic gates and some basic sequential logic systems such as flip-flops. There is not a great amount of detailed discussion of such circuits in these pages.

Microprocessor systems are made by combining registers, counters, arithmetic/logic units, encoders and decoders, memories, busses and other basic units. They operate by utilizing step-by-step programs.

A *bit slice* is a microprocessor that is made by using combinations of the above-mentioned logic units, which are manufactured as individual integrated circuits. Because this technology is available, it is possible to experiment with individual sections of microprocessors and microprocessor systems.

At first it might seem that this approach should culminate in a bit slice microprocessor experiment. There are two reasons why this was not done.

1. The timing complexities make it very difficult to construct a bit slice on a breadboard. Differences in propagation delay and transit time are sure to produce glitches and dead spots. The technical experience required to get the bugs out of a breadboarded bit slice is beyond the level of this entry-level book.

2. Bit slices are more difficult to program when compared to more conventional microprocessor systems. (This is partly due to the timing problems.) The designer must provide the program for the bit slice. No ready-made program is available.

Though the examples in this book mainly cover eight bit microprocessors, they can be utilized for sixteen and thirty-two bit microprocessors—the technology continues to grow, but the basics of dealing with it have changed little over the years.

The book starts with a general discussion of computers and microprocessor systems. That helps to keep the information in the remaining chapters in perspective.

Following that, each chapter deals with an important aspect of a microprocessor system.

The authors believe that it is not possible to understand microprocessors from a strict hardware approach, so there is some basic programming information included. Also, methods of troubleshooting microprocessor systems are included, because that is one of the most important reasons for studying microprocessors.

The experiments are innovative. It is assumed that a technician reading this book does not require a step-by-step procedure and a picture-book wiring diagram to accomplish the experiences that result from each experiment.

Chapter 1

Introducing the Computer
and the Microprocessor

In this chapter—

- An overall view of computers
- How the microprocessor got started
- The real purpose of the microprocessor
- The basic microprocessor system
- General descriptions of the units in a basic microprocessor system

Microprocessors have been described in many different ways. They have been compared with the brain and the heart of humans. Their operation has been likened to a switchboard, and to the nervous system in an animal. They have often been called micro*computers*.

The original purpose of the microprocessor was to *control memory*. That is what they were originally designed to do, and *that is what they do today!* Specifically, *a microprocessor is a component that implements memory.*

A very brief history of their development will serve to reinforce this concept. As the story goes, manufacturers of memories were able to get many memory addresses on a single integrated circuit. However, as the number of addresses increased, it became more and more difficult to wire them into logic circuits.

To reduce the wiring complexity, a Japanese manufacturer contacted two American manufacturers to make an integrated circuit device that could put information into memory and take it out of memory. Because this is very similar to the job of the *central processing unit* (CPU) in a large computer, the descriptive name for such a device is *microprocessor*.

The most significant feature of the microprocessor is that it is versatile. It can be used for various amounts of memory, and for various types of applications.

AN INTRODUCTION TO COMPUTERS

An overview of the computer will be helpful for a better understanding of microprocessors.

As you will see later in this chapter, there is not a great amount of difference between the operation of a computer system and the operation of a microprocessor system. That is why microprocessors have been called microcomputers. However, that name is confusing because the word microcomputer has come to mean computers that use microprocessors for their central processing unit. Figure 1-1 shows a basic block diagram for a computer.

Although we make a distinction in Fig. 1-1 between the basic computer, its power supply, and its peripherals, it is a common practice to refer to the complete system as being a computer.

The Basic Computer

The *central processing unit* (CPU) controls all of the system operations. It operates on the data, provides control signals for enabling and disabling the various sections, and determines the location of data and computer instructions as they are stored or retrieved.

The central processing unit also performs arithmetic and logic operations on the data. In a microcomputer, a microprocessor serves as the central processing unit.

Every computer has a *clock* for timing and sequencing the various operations. The clock produces a square wave. Various operations are performed during either the leading edge or the trailing edge of the clock pulses. In some computers there are two clock signals that are 180° out of phase. Both signals come from the same clock source.

A *crystal* provides the reference frequency from which the clock signal is derived.

Computers have an *input/output* (I/O) section, which permits the CPU to obtain data and instructions from the outside world and

2

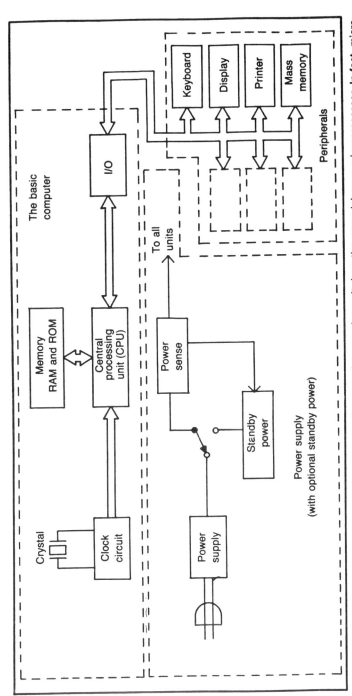

Fig. 1-1. In this simplified block diagram of a computer the central processing unit does the same job as a microprocessor. In fact, microprocessors are used as CPUs in personal computers.

3

to deliver its data and instructions to the outside world.

The Power Supply

All electronic systems, including computers and microprocessors, operate on dc power. Therefore, a power supply is included in the block diagram of Fig.1-1. It provides dc for all of the units in the microprocessor system. Most of the peripherals have their own supply.

Computer power supplies are very strictly regulated and filtered. An output of +5V is typical.

If there is a loss of power during the operation of a computer, some of the data in memory can be lost. It could be very expensive and time consuming to replace the data, so a *standby power system* is sometimes made available. In the event of ac power failure, it automatically switches on to prevent memory loss.

Because of the tight voltage regulation and the complicated electronic switching circuitry required for getting the standby power into operation before the memory is lost, computer power supplies are often more complicated than supplies in other systems.

Peripherals

The number of peripherals that a computer can accommodate is a measure of its versatility. You will sometimes hear the word *power* used to describe the number of peripherals. This is a misuse of the word power, but it is a common usage among small computer owners.

In the computer system there are *memories* that are used for storing the data and instructions. The same types of memories are used for microprocessor systems, and they will be discussed later in this chapter. (A more comprehensive discussion of memories is given later in the book.) Tapes, discs, and other systems permanently store a large amount of data and instruction. They are sometimes called *mass memories*.

A *keyboard* and *display* system are very important peripherals. They permit the user to put information into the computer and to read the output data from the computer. Printers are used to provide permanent (hard copy) readout information.

Depending upon the specific application desired, computers can be used to operate a wide variety of devices such as machinery, intrusion alarms, lights, speech equipment, etc. The versatility of the computers makes their variety of applications almost unlimited.

THE BASIC MICROPROCESSOR SYSTEM

A basic microprocessor system is very similar in operation to a computer, but it does not (usually) have the broad capabilities and versatility of a computer system. Very often, microprocessor systems are *dedicated*—that is, they are designed to perform a specific application rather than a wide variety of tasks.

As mentioned before, a microprocessor system may be used as the central processing unit of a computer. In modern microcomputers it is not uncommon to use one microprocessor system to operate another microprocessor system. This gives the microcomputer a much greater versatility.

Microprocessors will be introduced in this section and covered in greater detail in Chapter 2.

The Microprocessor

The microprocessor controls the flow of data and instructions into and out of memory. It also controls the flow of data and instruction into and out of the system. Arithmetic and logic operations are performed inside the microprocessor. Control signals originate in the microprocessor. Figure 1-2 shows Motorola's concept of a basic microprocessor system. It is based upon their 6800, which is a very popular example of a microprocessor. Other systems are similar, but different names or identifying numbers may be used for the sections.

The purpose of each block in the system is described in the following paragraphs.

Busses

A *bus* is a combination of wires or conductors that carries information in the form of codes made from binary 1's and 0's. In the 6800 system, a logic 1 is represented by +5V, and a logic 0 is represented by zero volts.

The *data bus* carries the information that is to be operated upon in the system. There are two kinds of numbers that appear on this bus: *data* and *instruction*. For example, if two numbers are to be added, these numbers and their sum would be called data. They would be delivered into and out of the system on the data bus.

Instructions are coded signals that tell the microprocessor to perform a certain task. For example, a certain coded number would be used to instruct the microprocessor to "take the next number you are given and put it into memory."

5

The *control bus* carries the control signals that choose various parts of the system for operations. For example, a control signal is used to select a particular memory when there are several memories to choose from.

The *address bus* carries the coded information that tells the location in memory where the microprocessor stores or retrieves data and instructions. The locations of the various blocks in Fig. 1-2 are

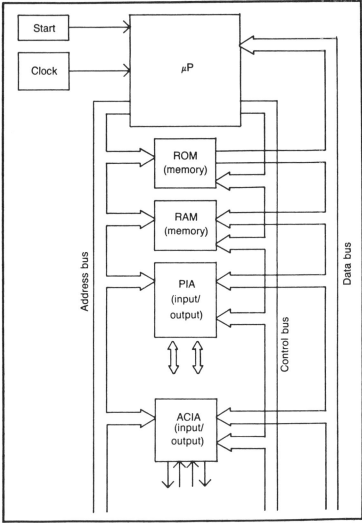

Fig. 1-2. This is the simplest microprocessor system. The PIA and ACIA are used for interfacing to the outside world.

also given as addresses. In other words, a microprocessor may consider these locations as being the same as memory locations.

Clocks

Clock signals are used in microprocessors for the same reasons they are used in computers. They move data and other logic signals in a pattern or sequence. Each clock transition produces a change or step in the operation. All microprocessors must have at least one clock signal for timing and operating the system.

Memories

Two kinds of memory are used in the basic system of Fig. 1-2. The *Random Access Memory* (RAM) is used for temporary storage of data and instructions. The microprocessor can put information in a RAM, and it can take information out of a RAM.

The *Read Only Memory* (ROM) has data and instructions stored permanently in its addresses. The microprocessor can take information out of a ROM; but, it cannot put new information into a ROM.

Interfacing

In the system of Fig. 1-2 there are two integrated circuit devices used for interfacing. They are the PIA and the ACIA. These abbreviations mean *Peripheral Interface Adapter* and *Asynchronous Communications Interface Adapter*. They connect the microprocessor to the outside world. The PIA allows the data and instruction to be interfaced eight **binary digits** (bits) at a time. This is called *parallel interfacing.*

The ACIA allows the data to be delivered or received one bit at a time. This is called *serial interfacing*, and it is ideal for connecting the microprocessor system to another through a telephone.

PROGRAMMED REVIEW

The chapter material is reviewed in this section. Some new material may also be introduced.

Read the question in block 1 carefully and select your answer. If choice (a) is your selection, go to block 8 as instructed. If you have chosen (b), go to block 12 as instructed.

If you make the wrong choice you will be told to re-read the question and then go to the proper block.

If your choice is correct there will be additional comments in the designated block. Then, a new question will be asked.

Be sure to review the chapter material for any questions you do not answer correctly!

Block 1

The original purpose of the microprocessor was to

(a) make computations. *Go to Block 8.*
(b) implement memory. *Go to Block 12.*

Block 2

Your answer to the question in Block 20 is not correct. Read the question again, then *go to Block 15.*

Block 3

The correct answer for the question in Block 19 is (a). Remember that a microprocessor is used for the central processing unit in personal computers. In the next chapter you will see how two numbers can be added in a microprocessor.

Here is your next question: Binary digits (1's and 0's) are used to make

(a) instruction codes. *Go to Block 16.*
(b) address codes. *Go to Block 9.*
(c) control codes. *Go to Block 27.*
(d) (All of these choices are correct.) *Go to Block 20.*

Block 4

The correct answer for the question in Block 15 is (a). In order to send parallel information over a telephone it would require eight telephone lines. That would be very expensive. With serial transmission, the binary codes are sent one bit at a time over one telephone line.

Here is your next question: In a computer, the section that connects the microprocessor with the outside world is

(a) R/L. *Go to Block 18.*
(b) I/O. *Go to Block 21.*

Block 5

The correct answer for the question in Block 24 is (c). The codes are called *data*. The directions for what to do with the data are called *instructions*. Both data and instruction codes are stored in memory. The combination of data and instructions which performs a task is called a *program*.

Here is your next question: Which of the following is a use of microprocessors in microcomputers?

(a) They are used in the same way as central processing units are used in large computers. *Go to Block 22.*
(b) They are used to produce timing signals. *Go to block 13.*

Block 6

Your answer to the question in Block 14 is not correct. Read the question again, then *go to Block 24.*

Block 7

Your answer to the question in Block 21 is not correct. Read the question again, then *go to Block 14.*

Block 8

Your answer to the question in Block 1 is not correct. Read the question again, then *go to Block 12.*

Block 9

Your answer to the question in Block 3 is not correct. Read the question again, then *go to Block 20.*

Block 10

Your answer to the question in Block 12 is not correct. Read the question again, then *go to Block 19.*

Block 11

Your answer to the question in Block 24 is not correct. Read the question again, then *go to Block 5.*

Block 12

The correct answer for the question in Block 1 is (b). INTEL was the company that made the first microprocessor. It was first fabricated early in the 1970 decade.

Here is your next question: Which of the following is the more important feature of a computer?

(a) Versatility. *Go to Block 19.*
(b) Low power supply voltage. *Go to Block 10.*

Block 13

Your answer to the question in Block 5 is not correct. Read the question again, then *go to Block 22.*

Block 14

The correct answer for the question in Block 21 is (a). Two things must be known in order to put information into (or take information out of) memory: the data (or the instruction) and the location. The location is called the address.

Here is your next question: A bus is made with

(a) a number of conductors. *Go to Block 24.*
(b) a number of semiconductors. *Go to Block 6.*

Block 15

The correct answer for the question in Block 20 is (b). Computers are usually sold with the keyboard, but it is still considered to be a peripheral.

Here is your next question: Which of the following types of interfacing would be best for connecting a microprocessor to a telephone line?
(a) Serial. *Go to Block 4.*
(b) Parallel. *Go to Block 26.*

Block 16

Your answer to the question in Block 3 is not correct. Read the question again, then *go to Block 20.*

Block 17

Your answer to the question in Block 24 is not correct. Read

the question again, then *go to Block 5.*

Block 18

Your answer to the question in Block 4 is not correct. Read the question again, then *go to Block 21.*

Block 19

The correct answer for the question in Block 12 is (a). Computers and microprocessors are both very versatile. In other words, the same system can be used for a wide variety of applications. This is one of their best features.

Here is your next question: In a computer, two numbers would be added in

(a) the CPU. *Go to Block 3.*
(b) the control circuit. *Go to Block 23.*

Block 20

The correct answer for the question in Block 3 is (d). All computers—including the largest and most modern—operate only with binary numbers. Combinations of these binary numbers are used to make up codes.

Here is your next question: Which of the following is an example of a computer peripheral?

(a) CPU. *Go to Block 2.*
(b) Keyboard. *Go to Block 15.*

Block 21

The correct answer to the question in Block 4 is (b). The Input/Output section of the computer is important because it determines which peripherals can be connected and the method of their connection. The same is true for microprocessor systems.

Here is your next question: Locations in memory where data is stored are called

(a) addresses. *Go to Block 14.*
(h) locales. *Go to Block 7.*

Block 22

The correct answer for the question in Block 5 is (a). The

microprocessor does the job of the central processing unit.

Here is your next question: Which of the following is not a type of bus in a basic microprocessor system?

(a) Arithmetic bus. *Go to Block 25.*
(b) Control bus. *Go to Block 28.*

Block 23

Your answer to the question in Block 19 is not correct. Read the question again, *then go to Block 3.*

Block 24

The correct answer for the question in Block 14 is (a). All computers and microprocessors have three types of busses: *data*, *address*, and *control*.

Here is your next question: On the data bus in a microprocessor system you will find codes for data and

(a) control signal. *Go to Block 11.*
(b) addresses. *Go to Block 17.*
(c) instructions. *Go to Block 5.*

Block 25

The correct answer for the question in Block 22 is (a). There is no such thing as an "arithmetic bus."

Here is your next question: In a computer the control signals come from the _____. *Go to Block 29.*

Block 26

Your answer to the question in Block 15 is not correct. Read the questions again, then *go to Block 4.*

Block 27

Your answer to the question in Block 3 is not correct. Read the question again, then *go to Block 20.*

Block 28

Your answer to the question in Block 22 is not correct. Read the question again, then *go to Block 25.*

Block 29

The correct answer for the question in Block 25 is CPU. You have now completed the programmed review.

EXPERIMENTS

Experiments in this book will give you insight into microprocessor operation.

Appendix A gives the specification sheets and pinouts in this book. There are many alternate integrated circuits available so you should consider the components specified in these experiments to be only guidelines. As an example, integrated circuits from the CMOS family can be used instead of the TTLs suggested in this book.

The Power Board

There are certain basic circuits that are used when performing most of the experiments in this book. Rather than build them each time you do an experiment, it is suggested that you construct all of them on a single board and save them.

For want of a better name, the board with these basic circuits will be called the *Power Board* in future chapters.

The Power Supply

We used a 9-volt lantern battery as a *power source* for performing the experiments. If you prefer, you can construct the bridge rectifier of Fig. 1-3. The choice of parts is not critical, but the diodes should be able to carry at least two amperes with a PIV of at least 50V.

The TTL family of integrated circuit logic is used in many of the experiments of this book. This *requires* that a regulated +5V supply be used. The output of the supply in Fig. 1-3 can be used as the input to the 5V 3-legged regulator shown later in this chapter.

Most of the TTL logic ICs have an equivalent CMOS integrated circuit. If you prefer to use the CMOS family you can use the 9V lantern battery without the regulator. However, you will still need the regulated supply for the memories and interface integrated circuits.

The suggested layout of the circuits on the power board is shown in Fig. 1-4. For convenience, the terminals for each circuit can be marked with labels.

14

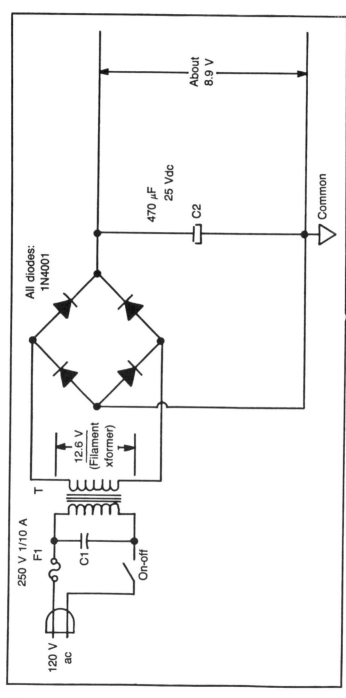

Fig. 1-3. This unregulated supply is used as the input of the 5-volt regulator (see Fig. 1-6). Two 6-volt batteries can be used in place of this circuit.

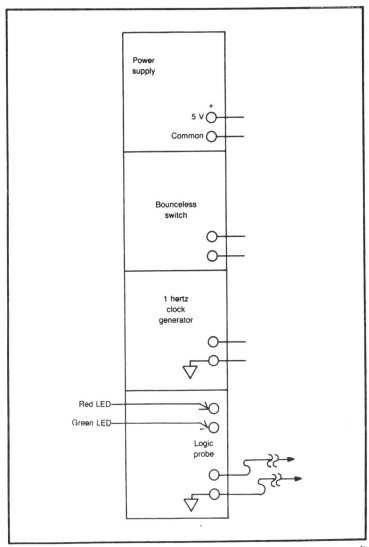

Fig. 1-4. Construct this power board in one corner of the experiment area. It can be used for the remaining experiments in the book.

Make your wiring neat so it is easy to troubleshoot. Avoid running wires over the top of integrated circuits because that makes it inconvenient to replace a defective IC on the board. Compare the wired boards in Fig. 1-5.

The circuits for the power board will now be discussed.

The +5 volt regulated supply is shown in Fig. 1-6. There are

15

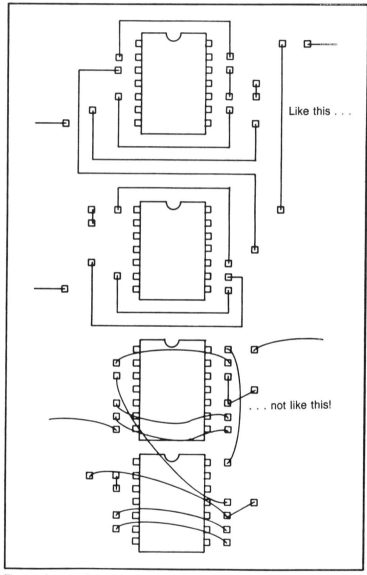

Fig. 1-5. A comparison of correctly - and incorrectly - wired boards. Which board would you rather troubleshoot?

a number of different three-legged regulators on the market so you can use a substitute without affecting the results of the experiments.

The output of the regulated supply is used to power the other circuits on this board, and it is also used for powering integrated

16

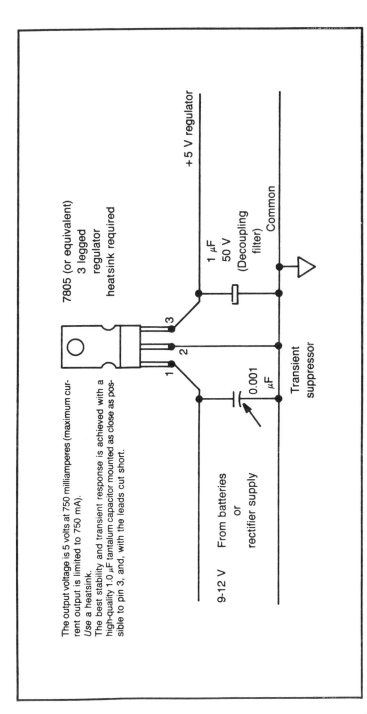

The output voltage is 5 volts at 750 milliamperes (maximum current output is limited to 750 mA).
Use a heatsink.
The best stability and transient response is achieved with a high-quality 1.0 µF tantalum capacitor mounted as close as possible to pin 3, and, with the leads cut short.

7805 (or equivalent)
3 legged
regulator
heatsink required

+ 5 V regulator

1 µF
50 V
(Decoupling
filter)

Common

3

2

1

0.001
µF

Transient
suppressor

9-12 V   From batteries
or
rectifier supply

Fig. 1-6 The regulated power supply circuit.

17

circuits built on another board. You can buy bus strips for these connections or you can use individual jumper wires for the +5V and common lines.

As a general rule, reducing the number of wires you need for connecting circuits will decrease the time you will spend troubleshooting.

The Bounceless Switch

The circuit for a bounceless switch is shown in Fig. 1-7. This circuit is needed for single-stepping sequential logic. All mechanical switches have contact bounce, and that will cause false triggering in a logic circuit.

The circuit of Fig. 1-7 is sometimes called an *R-S flip-flop*. A single operation of the push-button switch will cause a single output pulse. Remember that TTL logic transitions usually occur on the *trailing* edge of the pulse, so the transition will occur when you release the push button. (If you are using CMOS logic the transition occurs on the *leading* edge in most cases.)

The LED across the output of the flip-flop should be *on* while the button is pushed and *off* when it is released.

The Clock

The circuit for the clock is shown in Fig. 1-8. There are many different versions of this clock and you should learn to recognize clock circuits in synchronous logic and microprocessor systems. Without the clock pulse the system will not work.

In terms of troubleshooting any microprocessor system, there are two important places to start:

First - the power supply

Second - the clock pulse

The clock pulses for the circuit in Fig. 1-8 should occur at a rate of (approximately) one pulse per second, so the LED should flash on and off at that rate. You may find this too slow for some circuits, so you can adjust the rate with the variable resistor.

There are other popular clock circuits used for sequential logic systems. The 555 timer is popular. However, if you use a 555 timer to obtain the clock pulse you will have to wire another integrated circuit and that increases the power drain on the supply.

We recommend the circuit in Fig. 1-8 for another reason. It is the one you will see more often in commercial systems.

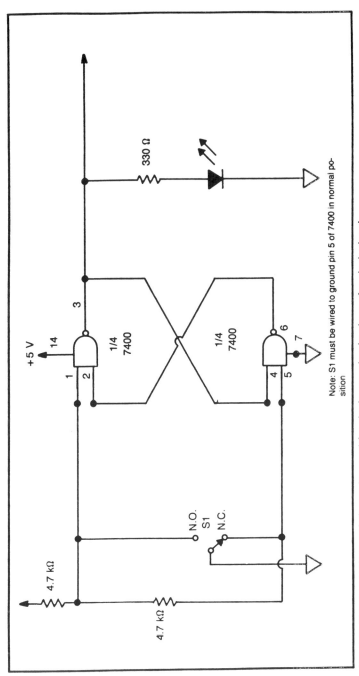

Fig. 1-7. This bounceless switch can be used for single stepping in place of a clock signal.

19

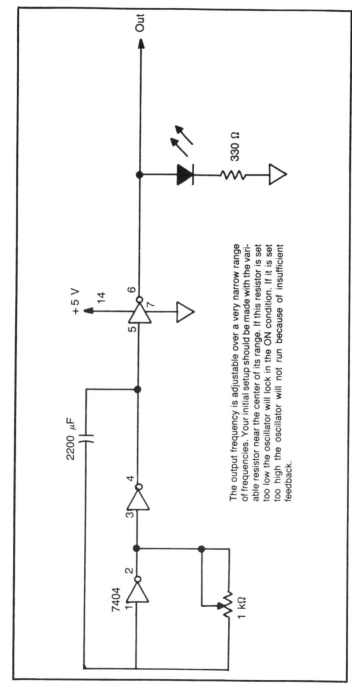

The output frequency is adjustable over a very narrow range of frequencies. Your initial setup should be made with the variable resistor near the center of its range. If this resistor is set too low the oscillator will lock in the ON condition. If it is set too high the oscillator will not run because of insufficient feedback.

Fig. 1-8. All microprocessor systems and synchronous logic systems *require* a clock signal. This popular clock signal can be used for low frequency pulses.

The Logic Probe

Figure 1-9 shows the circuit for a simple logic probe. This circuit also uses inverters in the hex inverter integrated circuits. There is a disadvantage of having the LEDs on a separate board from where you are troubleshooting but after a few measurements you will become accustomed to using the probe this way.

By using a green LED for common (logic 0) and a red LED for a +5V (logic 1) you will greatly reduce the likelihood of measurement error.

As soon as you have constructed the logic probe, use it to test the various circuits on the power board. Observe that the red and green LEDs will flash when you are checking the output of the clock. If the clock operates at a high frequency they will both appear to be on at all times.

SELF TEST

(Answers at the end of the chapter)

1. Microprocessors are
    (a) only used for calculations.
    (b) a form of memory.
    (c) clocked by a sine wave.
    (d) used for implementing memory.
2. The frequency of timing circuits in microprocessor systems is determined by
    (a) a crystal.
    (b) an R-C circuit.
    (c) an R-L circuit.
    (d) an L-C circuit.
3. You would expect to find instruction codes
    (a) on the control bus.
    (b) on the data bus.
    (c) on the address bus.
    (d) (None of these choices is correct.)
4. A combination of conductors used to carry information in the form of binary codes is called
    (a) a transmission line.
    (b) a carrier.
    (c) a transmitter.
    (d) (None of these choices is correct.)

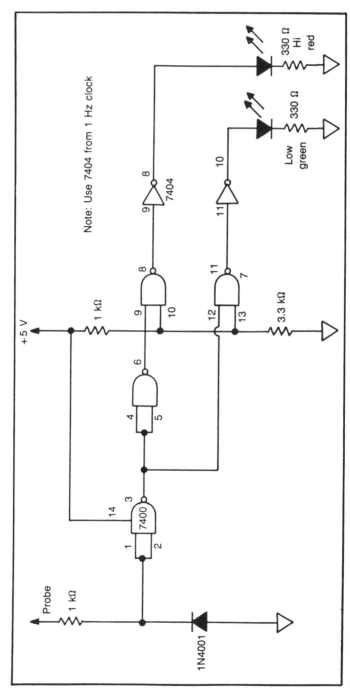

Fig. 1-9. There are many types of logic probes used for troubleshooting. This one is not only useful for the experiments but you can construct one for your work.

22

5. Memories are used to store data and
   (a) control signals.
   (b) addresses.
   (c) instructions.
   (d) clock signals.
6. Compared with the power supply circuitry in a radio, the supply used in a computer system is usually
   (a) more complicated.
   (b) less complicated.
7. An audio cassette is used with a certain computer to store data and instructions. This is an example of
   (a) RAM.
   (b) ROM.
   (c) mass memory.
   (d) EAROM.
8. Hard copy from a computer is obtained with a
   (a) camera.
   (b) telescope.
   (c) tape.
   (d) printer.
9. A microprocessor that is designed to perform one special job, or only a few jobs, is said to be
   (a) special case.
   (b) dedicated.
   (c) front loaded.
   (d) scripted.
10. A unit that is used with a computer but is located outside the computer is called
    (a) an add-on.
    (b) a peripheral.
    (c) an after product.
    (d) a clock.

Answers to the Self Test

1. (d)
2. (a)
3. (b)
4. (d) The correct answer is Bus.
5. (c)

6. (a)
7. (c)
8. (d)
9. (b)
10. (b)

Chapter 2

How Does the Microprocessor Work?

In this chapter—

- An example of an *instruction set*
- A program for adding two numbers
- How the microprocessor implements the program
- The purpose of a monitor
- The need for a CCR

In Chapter 1 it was stated that the purpose of a microprocessor is to implement memory. A good way to understand how it accomplishes this purpose is to follow a simple program into, through, and out of the microprocessor.

To do this, a simple 4-bit microprocessor will be used. It is shown in Fig. 2-1. Since there are only four bits available for instructing the microprocessor, it follows that there are only 2^4 actual instructions.

Every manufacturer of microprocessors provides an instruction set with its product. It lists all of the things which you can make the microprocessor do. As stated before, a 4-bit microprocessor has only 16 instructions. For the microprocessor used in this chapter, the instruction set is shown later. This instruction set will be used for writing the program to be discussed.

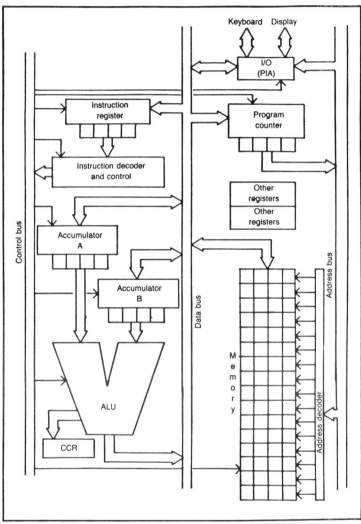

Fig. 2-1. This microprocessor block diagram will be used for the discussion on programs.

FEATURES OF THE MICROPROCESSOR

The following sections in Fig. 2-1 are located outside the microprocessor:

Keyboard. The keyboard is used for entering the program.

Display. The display shows the program numbers as they are entered. It also shows the answer to the problem.

I/O. All information and control signals pass from the

microprocessor system to and from the outside world through the I/O. In this case it is called a PIA (peripheral interface adapter). That is a Motorola name. In systems designed by other companies it has other names. One example is PIO (peripheral input/output).

Memory. The memory shown is RAM. It will be used to store the program. Note that there are two connections to the RAM. One comes from the *control bus* and *address decoder*. This input selects the location of information. The other is from the *data bus*. The data and the instructions move along this input.

The program counter is made with flip flops. It sequences through the RAM addresses as the program is run.

Since there are only 16 addresses in the RAM, the program counter will always start with 0000 (for address 0) in this simple system. In more sophisticated systems the user decides where the program will be stored in memory.

Remember that the RAM stores both data and instructions. When an instruction is implemented it is taken from the RAM to the *instruction decoder*. To get there it is first stored in the *instruction register*.

The ALU does the arithmetic work. It is also used to perform logic operations. The two numbers into the ALU are stored in the accumulators, and the result of the operation is also stored in an accumulator.

A *Condition Code Register* (CCR) is connected directly to the ALU. It stores information directly related to ALU operations. For example, if two numbers are added and there is a carry, it will be shown by the CCR. Also, a code for either a positive or negative answer is stored as a binary number in the CCR.

THE INSTRUCTION SET

Table 2-1 is the instruction set. The binary numbers are called OP codes. They tell the microprocessor what to do. Some are for arithmetic operations and some are for logic operations. The balance are for microprocessor internal instruction.

The mnemonics are useful for remembering the instructions. They are easier to remember than the binary OP codes.

THE PROGRAM

A program is a combination of op codes and data that accomplishes a certain task.

The first step in the program must be an op code. It wouldn't

Table 2-1. Instruction Set for Microprocessor of Fig. 2-1.

OP Code	Description	Mnemonic
0000	Load Accumulator A with the next number in the program M—►A *	LDAA
0001	Load Accumulator B with the next number in the program M—►B	LDAB
0010	Compare Accumulators A − B	COM
0011	Add Accumulators and store the sum in Accumulator A. A + B—►A	ABA
0100	Clear the memory location 00—►M	CLR
0101	Store Accumulator A in the location designated by the next number in the program A—►M	STAA
0110	Store Accumulator B in the location designated by the next number in the program B—►M	STAB
0111	Subtract Accumulators and store the difference in Accumulator B. A − B—►A	SBA
1000	Transfer information in Accumulator A to Accumulator B. A—►B	TAB
1001	Transfer information in Accumulator B to Accumulator A. B—►A	TBA
1010	Display location	D
1011	AND accumulators, store result in Accumulator A. A × M—►A	ANDA
1100	OR Accumulators, store result in Accumulator B A + B—►B.	ORA
1101	Increment the contents of Accumulator A. Store result in Accumulator A. A + 1—►A	INCA
1110	Decrement the contents of Accumulator A. Store the result in Accumulator A. A − 1—►A	DECA
1111	Stop the Program	WAI

*The M indicates that the number to be loaded is in memory

be possible to start with data because the microprocessor wouldn't know what to do with it.

The program that will be used as an example is a simple addition of two numbers (2 and 3). In order to get the ALU to add these numbers they must first be stored in the accumulators. It takes six steps to accomplish this.

Step 1. An op code tells the microprocessor to load the next number in the program into Accumulator A. (LDAA)

Step 2. The next number in the program is the binary equivalent of 2. The microprocessor will put this number into Accumulator A.

Step 3. An op code tells the microprocessor to load the next number in the program into Accumulator B. (LDAB)

Step 4. The next number in the program is the binary equivalent of 3. The microprocessor will put this number into Accumulator B.

Step 5. An op code tells the microprocessor to add the numbers that have been stored in the accumulators. After the numbers have been added, the answer, or *sum,* is stored in Accumulator A.

In this simple program the answer is left in the accumulator. In other programs the answer may be sent to a display, stored in memory, added to another number, or any number of choices that can be selected from the Instruction Set.

Step 6. The program has been run, so, the microprocessor must be told to stop. The instruction WAI (for wait) puts the microprocessor in a standby condition. This step is essential to the program. Without it the program counter would step through the random numbers in memory and the microprocessor would try to implement those numbers. The desired answer would be lost.

Table 2-2 shows the program. Before proceeding with this chapter, write your own program for subtracting 5 (binary 0101) from six (0110). Store the answer in the accumulator. Space is provided for your program in Table 2-2.

After you have written your program, compare it with the one given at the end of this chapter.

Implementing the Program

Once the program is written, the next step is to load it into memory. Then, the microprocessor will take each step and implement it.

The keyboard is used to enter the program. The microprocessor stores the entered program into memory.

The first thing that must be done is to clear all of the registers, accumulators, and the program counter. This would be a tedious step-by-step procedure, so manufacturers provide an integrated circuit to automatically take care of this and other routine jobs. This IC is called a *monitor.*

Figure 2-2 shows the system with the memory loaded and all of the registers, accumulators, and the program counter cleared (set to 0000). The program is now ready to run. Note that the program counter is set to 0000. That is the binary code for address 0 in memory. As the program continues, the program counter will step through the addresses.

It is important to note that the program in memory is loaded

Table 2-2. Program for Adding 2 and 3, and Form for Your Own Program.

Step	Data or Instruction?	Program	Description
1	Op Code	0000	LDAA
2	Data	0010	Binary Equivalent of 2
3	Op Code	0001	LDAB
4	Data	0011	Binary Equivalent of 3
5	Op Code	0011	ABA
6	Op Code	1111	WAI

Your Program:

into addresses 0 through 5. The numbers in addresses 6 through 15 are random, or they may be left over from a previous program. It is obvious that the last step in our program must halt the microprocessor. Otherwise, it would continue to step through the remaining numbers trying to implement them as part of the program.

In the next six figures, the program is stepped through the microprocessor. As you follow the program, refer back to the meanings of each step, which were given earlier in the chapter.

The program is started by operating a key called "GO," or "START," or some other description.

Figure 2-3. The program counter is now started. It is set to select address 0000 (or, address 0). The op code in that address is delivered to the Instruction Register and then to the Instruction Decoder and Control. It sends a control signal to Accumulator A to open it in preparation for the next number that appears on the data bus.

Figure 2-4. The program counter sequences to 0001. That selects Address 1 which has the number to be stored in Accumulator A. The number in Address 1 is then delivered to Accumulator A.

Figure 2-5. The program counter sequences to Address 0010

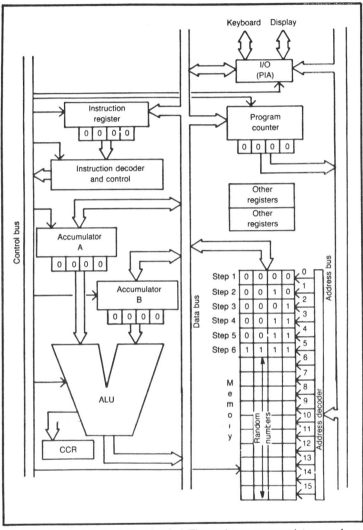

Fig. 2-2. The memory has been loaded. The registers, accumulators, and program counter have been reset to zero. The microprocessor is now ready to run the program.

(Address 2). This is an op code that goes to the instruction register —then to the instruction decoder. The decoder opens Accumulator B in preparation for the next number to come into the data bus.

Figure 2-6. The program counter sequences to the next address which is 0011 (3). The number in that address is delivered to Accumulator B.

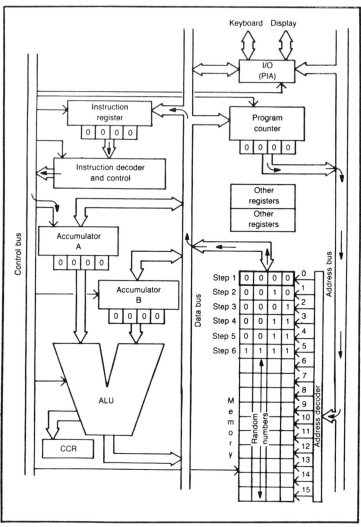

Fig. 2-3. The microprocessor selects the first memory address. The data in that location is an instruction. It is decoded in the instruction register.

So far, the number 2 has been stored in Accumulator A and the number 3 has been stored in Accumulator B.

Figure 2-7. The program counter sequences to address 0100 (4). The op code in that address is sent to the instruction register and then to the decoder. The decoded signal tells the ALU to add the two numbers in the accumulators and store the sum in Accumulator A. The dotted arrows show the path of the answer.

32

Figure 2-8. The program is ended. The program counter sequences to 0101. This op code goes to the instruction register and decoder. The instruction stops the program counter. The microprocessor is waiting for a new instruction.

PROGRAMMED REVIEW

(See the instructions for the Programmed Review in Chapter 1.)

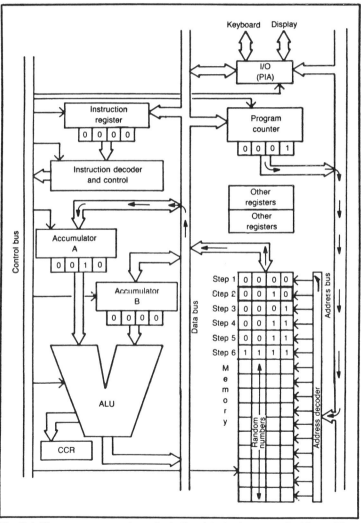

Fig. 2-4. The program counter selects the data in the second address. That data is delivered to Accumulator A.

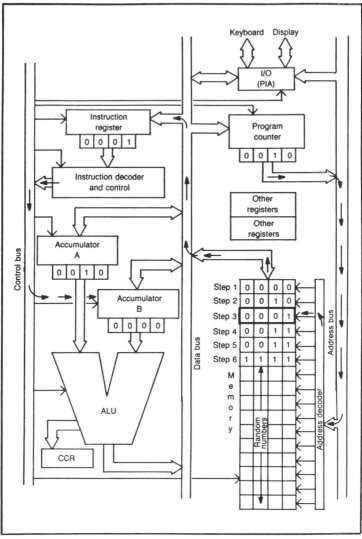

Fig. 2-5. The program counter selects the third address. The instruction in that address is decoded in the instruction register.

Block 1

The address of the next step in the program is in

(a) the instruction register. *Go to Block 8.*
(b) the program counter. *Go to Block 12.*

Block 2

Your answer to the question in Block 12 is not correct. Read the question again, then *go to Block 24.*

Block 3

The correct answer for the question in Block 11 is (a). The CCR

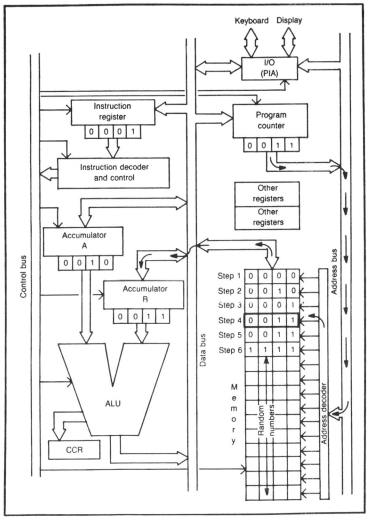

Fig. 2-6. The program counter selects the fourth address. The data in that address is loaded into Accumulator B.

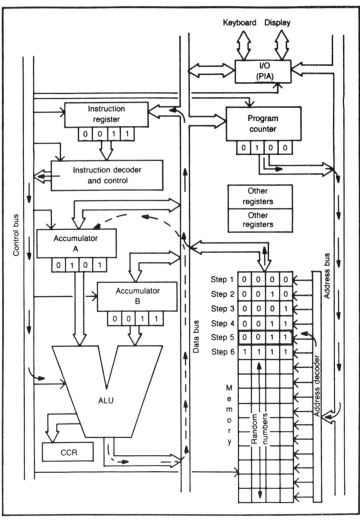

Fig. 2-7. The program counter selects the fifth address. The instruction in that address is decoded by the instruction register. The ALU adds the accumulator numbers and stores the result in Accumulator A.

is connected to the ALU. In many microprocessor *systems* the user can look at the numbers stored in the CCR. Here is your next question: What determines if an answer is positive or negative after a subtraction has been performed?

(a) A binary number in the CCR. *Go to Block 22.*
(b) A binary number in the instruction decoder. *Go to Block 14.*

36

Block 4

The correct answer for the question in Block 9 is (a).

Here is your next question: The Motorola name for the microprocessor system I/O is

(a) PIA. *Go to Block 18.*
(b) ABL. *Go to Block 13.*

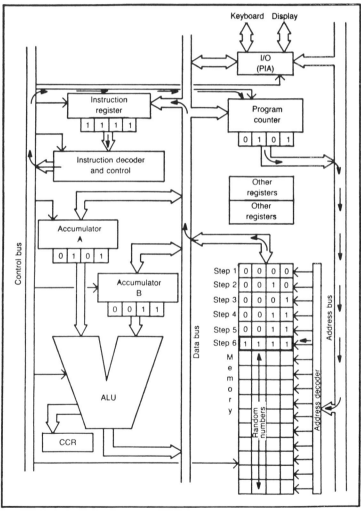

Fig. 2-8. The program counter selects the sixth address. The op code in this address is decoded in the instruction register. The instruction decoder and control stops the program counter and therefore stops the program.

Block 5

Your answer to the question in Block 22 is not correct. Read the question again, then *go to Block 29.*

Block 6

Your answer to the question in Block 11 is not correct. Read the question again, then *go to Block 3.*

Block 7

Your answer to the question in Block 29 is not correct. Read the question again, then *go to Block 20.*

Block 8

Your answer to the question in Block 1 is not correct. Read the question again, then *go to Block 12.*

Block 9

The correct answer for the question in Block 23 is (b). The instruction codes are dependent upon the way the microprocessor is designed.

Here is your next question: How many instructions can an 8-bit microprocessor have?

(a) $2^8 = 256.$ *Go to Block 4.*
(b) $2 \times 8 = 16.$ *Go to block 26.*

Block 10

Your answer to the question in Block 24 is not correct. Read the question again, then *go to Block 16.*

Block 11

The correct answer for the question in Block 28 is (b). You will construct a similar counter in the experiment section of this chapter.

Here is your next question: In a certain program two numbers are added and there is a 1 carried. Where is the 1 stored?

(a) In the CCR. *Go to Block 3.*
(b) In the instruction decoder. *Go to Block 6.*

Block 12

The correct answer for the question in Block 1 is (b). The program counter is sequenced by clock pulses.

Here is your next question: The program is loaded into memory by using the

(a) control bus. *Go to Block 2.*
(b) keyboard. *Go to Block 24.*

Block 13

Your answer to the question in Block 4 is not correct. Read the question again, then *go to Block 18.*

Block 14

Your answer to the question in Block 3 is not correct. Read the question again, then *go to Block 22.*

Block 15

Your answer to the question in Block 29 is not correct. Read the question again, then go to Block 20.

Block 16

The correct answer for the question in Block 24 is (b). You have to start the program by telling the microprocessor what to do first.

Here is your next question: The CCR is

(a) a charge-coupled register. *Go to Block 19.*
(b) a condition code register. *Go to Block 23.*

Block 17

Your answer to the question in Block 23 is not correct. Read the question again, then *go to Block 9.*

Block 18

The correct answer for the question in Block 4 is (a). (There is no ABL code for the microprocessor discussed in this chapter.)

Here is your next question: The program is stored in

(a) a RAM. *Go to Block 28.*

(b) a register. *Go to Block 25.*

Block 19

Your answer to the question in Block 16 is not correct. Read the question again, then *go to Block 23.*

Block 20

The correct answer for the question in Block 29 is (c). Check this in the instruction set.

Here is your next question: Which of the following is the last instruction in a program?

(a) WAI. *Go to Block 27.*
(b) STAB. *Go to Block 30.*

Block 21

Your answer to the question in Block 28 is not correct. Read the question again, then *go to Block 11.*

Block 22

The correct answer for the question in Block 3 is (a).

Here is your next question: Easy-to-remember abbreviations (such as LDAA and ABA) are called

(a) MNEMONICS. *Go to Block 29.*
(b) PSEUDONYMS. *Go to Block 5.*

Block 23

The correct answer for the question in Block 16 is (b). The condition code register holds bits that determine whether the answer to a problem solved by the ALU is positive or negative. Another example of a bit in the CCR is one that determines if there is a carry after addition.

Here is your next question: The instruction codes of a microprocessor are

(a) developed by the user for his particular needs. *Go to Block 17.*
(b) provided by the microprocessor manufacturer. *Go to Block 9.*

Block 24

The correct answer for the question in Block 12 is (b). There are other ways of entering programs—such as switches—but this is the only correct answer for the choices given.

Here is your next question: The first step in a program must be

(a) data. *Go to Block 10.*
(b) an op code. *Go to Block 16.*

Block 25

Your answer to the question in Block 18 is not correct. Read the question again, then *go to Block 28.*

Block 26

Your answer to the question in Block 9 is not correct. Read the question again, then *go to Block 4.*

Block 27

The correct answer for the question in Block 20 is (a). When the WAI op code is implemented the microprocessor goes into a standby condition. The program counter stops sequencing.

Here is your next question: A similarity between a microprocessor and computer is that they both implement _____. *Go to Block 31.*

Block 28

The correct answer for the question in Block 18 is (a). The size of the RAM is a limiting factor as to how complex a program can be.

Here is your next question: The program counter is made with

(a) a decoder. *Go to Block 21.*
(b) flip-flops. *Go to Block 11.*

Block 29

The correct answer for the question in Block 22 is (a). Mnemonics make it easier for the programmer to remember the instructions.

Here is your next question: What is the meaning of A + B →

A in the list of instructions?

(a) A + B is greater than A. *Go to Block 15.*
(b) A or B is equal to A. *Go to Block 7.*
(c) Add the numbers in Accumulators A and B, then store the answer in A. *Go to Block 20.*

Block 30

Your answer to the question in block 20 is not correct. Read the question again, then *go to Block 27.*

Block 31

The correct answer to the question in Block 27 is *memory.* You have now completed the programmed review.

EXPERIMENTS

In the experiments for this chapter you construct some of the basic circuits used in a microprocessor.

The Counter

From your study of this chapter you know that a program counter is used to sequence the addresses in memory where the program has been stored. Figure 2-9 shows two types of counters.

Figure 2-9 (A) shows a ripple counter made with flip-flops. This is an asynchronous counter. It is sometimes called a ripple counter because when a number changes each flip-flop changes in sequence. In other words the change ripples through the flip-flop as they change one at a time.

Figure 2-9 (B) shows a synchronous counter. Note that the clock goes to all of the flip-flops at the same time. So, for each clock pulse, all of the numbers change at the same instant. The synchronous counter is much faster than a ripple-through counter but it has a disadvantage of requiring more power for changing the count. Microprocessors use synchronous counters. Construct the synchronous counter.

When the counter is running (with a 1 Hz clock pulse input) stop the count by switching J and K from 1 to 0. The counter is **never** stopped by stopping the clock. Note also that it is necessary to switch all of the J-K terminals to stop the count. (With a ripple counter it is only necessary to switch the J and K in the first flip-flop to stop the count.)

The counter can be programmed to count to a certain number and then stop. Figure 2-10 shows how the counter can be made to count from 0 to 5 and then stop.

(Programmable counters are available. An example is the 74191. They can be easily set to count to any number.)

Observe that the count is stopped in this ripple counter by stopping the first flip-flop. This is done by delivering a logic 0 to the J_1 and K_1.

SELF TEST

(Answers at the end of the chapter)

1. Which of the following is an integrated circuit controlling device that takes care of tedious routine jobs like clearing all registers in a microprocessor?

    (a) Op amp
    (b) Crow bar
    (c) Monitor
    (d) Sequencer

2. Answers to problems solved by the ALU are first stored

    (a) in an accumulator or register.
    (b) in the CCR.
    (c) in the instruction register.
    (d) in other registers.

3. The Motorola PIA is an example of

    (a) a clock.
    (b) a monitor.
    (c) a CCR.
    (d) an I/O device.

4. A binary code is used to select an address in memory. This is accomplished with

    (a) an address accumulator.

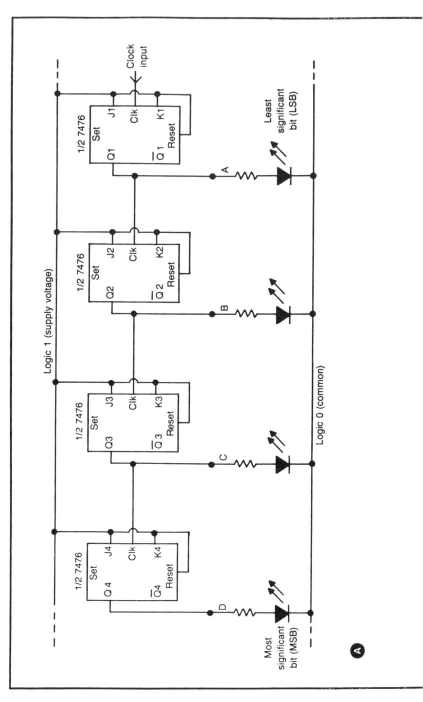

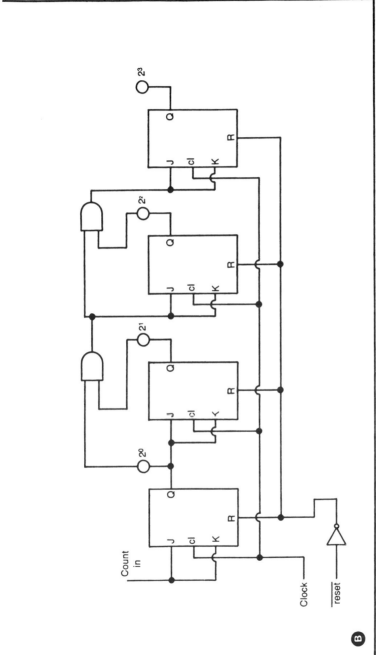

Fig. 2-9. This ripple counter will sequence from 0000 through 1111 and then start over. Note that the clock pulse is on the right. This permits us to draw the counter so that the least significant bit is on the right and the most significant bit is on the left.

45

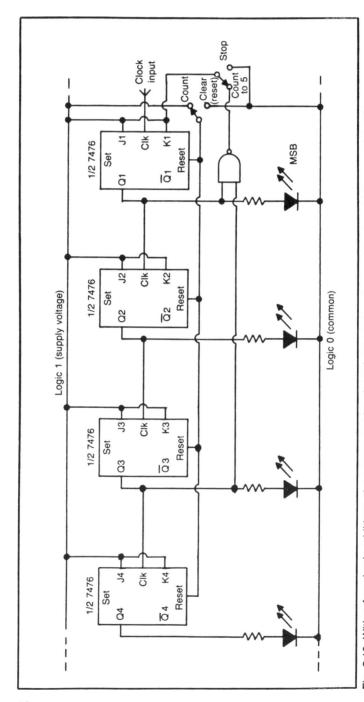

Fig. 2-10. With a few simple additions, the counter can be cleared, stopped, and caused to stop at binary number 101.

(b) an address register.
(c) an address decoder.
(d) the CCR.

5. Which of the following is NOT a type of bus *required* by a microprocessor?

   (a) Keyboard
   (b) Control
   (c) Data
   (d) Address

6. A microprocessor instruction is supplied to the user by

   (a) the on-site programmer who works for the user.
   (b) the manufacturer.
   (c) the sales department of the user.
   (d) an independent publisher.

7. When the microprocessor is performing a task, addresses of the program come from

   (a) the ALU.
   (b) the CCR.
   (c) the instruction register.
   (d) the program counter.

8. How many different instructions can be written by using four binary bits in a code?

   (a) 4
   (b) 16
   (c) 64
   (d) 256

9. Instead of the PIA used by Motorola, another company may use a

   (a) PEA.
   (b) PIO.
   (c) POA.
   (d) PCC.

10. The real purpose of a microprocessor is to

   (a) control machinery.
   (b) operate computers.
   (c) add numbers.
   (d) implement memory.

Answers to the Self Test

   1. (c)
   2. (a)
   3. (d)
   4. (c)
   5. (a)
   6. (b)
   7. (d)
   8. (b)
   9. (b)
   10. (d)

Chapter 3

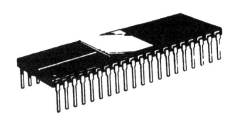

The Microprocessor
from the Outside

In this chapter—

- Abbreviations for pin identifications
- The job of each pin
- Some troubleshooting ideas
- The dedicated microprocessor
- Repairing surface mounting boards

So far, we have looked at the microprocessor from the inside. We have seen how a program is stepped through the various sections of the microprocessor and how the microprocessor makes use of memory to store the program and to implement the program.

Now we are going to look at the microprocessor from the outside. In other words, this is the view as seen by the technician. We will use the Motorola 6800 microprocessor first, then compare it to the 6802. One difference between the two microprocessors is that the 6800 requires an external clock circuit and the 6802 has that clock circuit built in.

We are going to discuss these microprocessors pin by pin and identify their functions.

We will also make some comment about troubleshooting the microprocessor as we proceed. Refer to Fig. 3-1 for each of the following pin discussions.

Pin 8 is marked VCC. This microprocessor operates from a

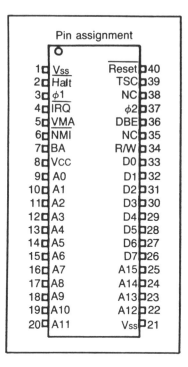

Pin assignment

1	Vss	Reset	40
2	Halt	TSC	39
3	φ1	NC	38
4	IRQ	φ2	37
5	VMA	DBE	36
6	NMI	NC	35
7	BA	R/W	34
8	VCC	D0	33
9	A0	D1	32
10	A1	D2	31
11	A2	D3	30
12	A3	D4	29
13	A4	D5	28
14	A5	D6	27
15	A6	D7	26
16	A7	A15	25
17	A8	A14	24
18	A9	A13	23
19	A10	A12	22
20	A11	Vss	21

Fig. 3-1. The pinouts for the 6800 microprocessor are shown here.

regulated +5V power supply. So, whenever you see VCC in the 6802 (or 6800) circuit you can expect a 5V potential at that point. In this case VCC is the power supply connection for the microprocessor. You should expect to see the dc voltage at this pin when the microprocessor is in operation.

As with other electronic devices, the microprocessor will not operate without a supply voltage. The abbreviation VCC is used in bipolar transistor terminology. It is a collector voltage. More specifically, it is a collector *supply* voltage.

Pins 1 and 21 are marked VSS. That symbolism is normally used with Field Effect Transistor (FET) technology. The SS refers to an FET source grounding location.

At pins 1 and 21 you would expect to see zero volts when the microprocessor is in operation. Note that the manufacturer has put VSS on both sides of the integrated circuit. That makes the design of printed circuit boards much easier. A big problem in printed circuit board design is to make them so that the conductors aren't crossing over each other. It might be easy to get VSS on one side but impossible on the other side. So, the manufacturer has simplified the mounting of this device on printed circuit boards.

50

Pin 2 is marked $\overline{\text{HALT}}$. Any time you see a line over the top of a microprocessor or digital symbol, it means NOT or NO. That means there must be a logic 1 into this pin to prevent the microprocessor from stopping what it is doing. During the normal operation of the microprocessor then, you will measure a +5V at this terminal.

The manufacturer uses the term *active low* for symbols that have an overbar like the one on pin 2. It means that if you want to halt the microprocessor you must put a logic $\underline{0}$ (ground) into that pin.

When the microprocessor receives a $\overline{\text{HALT}}$ signal (0V on pin 2) it will finish executing the instruction it is working on, then it will wait for further instruction. However, it will store an interrupt signal from the outside world while it is in the HALT condition.

An important troubleshooting technique is called *single stepping*. (This technique is also used for stepping through a program to look for errors.)

Single stepping permits the technician to check signals on the microprocessor pins as each step of a program is executed—one step at a time.

Single stepping occurs by delivering zero-volt pulses to the $\overline{\text{HALT}}$ pin (pin 2) of the microprocessor.

Pin 3 is a Phase I connection. The clock signal for this microprocessor consists of two separate voltages. They are 180° out of phase pulses as shown in Fig. 3-2. This arrangement makes it possible for the microprocessor to operate more quickly.

One phase is used to get the microprocessor ready to do something and the other phase is used when it is doing its job.

If you will look on the other side of the microprocessor to pin 37, you will see the Phase II connection. Looking at the signals on these two pins with a dual trace oscilloscope, you should be able to see the signals as illustrated in Fig. 3-2.

Note that the two clock signals are *nested*. In other words, one of the signals is OFF completely before the other signal is ON. This arrangement assures that one operation is completely finished before another starts.

The microprocessor will not operate unless both clock signals are present. They are actually derived from a crystal-controlled clock. As mentioned before, the clock circuitry is built into the microprocessor, but the crystal itself is outboard.

Not all microprocessors use two separate clock signals. Some of them use simply a single phase clock. However, all microprocessors will have at least one clock signal.

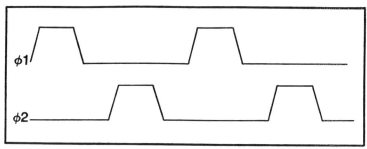

Fig. 3-2. Two clock signals are required by the 6800 microprocessor. These clock signals are nested, as shown here so that there is no possibility that they can be at logic 1 simultaneously.

If you are troubleshooting a microprocessor circuit, the place to start is by measuring the power supply voltages and the clock signals! If they are not present, the microprocessor cannot operate and the trouble is not inside the microprocessor.

It is important to remember that the signals and voltages must actually be delivered to the microprocessor. So, they must be measured at the pins at the point where they enter the microprocessor case. Don't be guilty of just measuring these at some convenient point on the printed circuit board. The signal or voltage can be present at that point and still not be getting into the microprocessor due to a faulty connection.

Pin 4 is marked $\overline{IRQ}$. This means the pin is used for a *NOT interrupt request*. In other words, it is an *active low interrupt request.*

An interrupt request comes from outside the microprocessor. It is simply a way of telling the microprocessor that there is information available and it should be used as soon as the microprocessor can get to it.

The microprocessor will finish what it is doing first, then it will store the information in its registers and accumulators before it goes to work on the interrupt signal.

It is possible to program the microprocessor to ignore the interrupt. This is called *masking* the interrupt.

Pin 5 is called VMA. It means *valid memory address.* You have to distinguish between data pulses and undesired noise pulses in a microprocessor system. The VMA pin is used by the microprocessor to tell the memory that this is really a valid signal.

Pin 6 is identified as $\overline{NMI}$. The symbol means *nonmaskable interrupt.* It is another active low pin.

This is an interrupt that has to be taken seriously by the microprocessor. It cannot shut out this interrupt signal, and it is not going to go away. This signal means business. By contrast, another type of interrupt tells the microprocessor that it has information to be serviced whenever it is convenient for the microprocessor. That's the type of interrupt that is delivered to pin 4.

When there is an interrupt signal on pin 6 the microprocessor finishes its present instructional step. Then it saves the information in the accumulators and registers before it starts to service the interrupt.

Pin 7 is marked BA which stands for bus available. This pin is like a streetcar conductor or traffic cop. You wouldn't want to try to send signals into the microprocessor and out of the microprocessor both at the same time. The BA pin will prevent this.

You will demonstrate in one of your experiments how this two-way bus system operates. It permits a single bus to be used for signals that are going out and also for signals that are coming into the microprocessor system.

Pins 9 through 20 and **pins 22 through 25** are marked with the letter A. There are a total of 16 pins marked with this identification. The A stands for address. You will remember in the simple program that we discussed in Chapter 2 that the microprocessor had to go to various addresses in the memory in order to implement the program. The signal for locating the address would be on one of these busses.

Because of the fact that the lines can carry only ones and zeros, it follows that with 16 address line you can have 2^{16} or 65,536 possible codes that you can utilize with these 16 pins. Another way of saying this is that the microprocessor can direct signals to 65,536 different locations. The address is the locations where the microprocessor is sending a signal to or receiving a signal from.

In microprocessor language, the number 65,536 is called 64K (64,000). Usually a person studying this for the first time wants to know why they call it 64K instead of 65,536. The reason for this is that the power of two that is nearest to 1000 decimal is 2^{10}, which equals 1024. Among early computer engineers, who were by necessity very familiar with binary numbers, this number came to be known as a K, for kilo-. It is now easy to see that 64 x 1024 or 64K is equal to 65,536.

The system in which this microprocessor is used is called *memory mapped*. The way it works is that the microprocessor considers any place that a signal is going to or coming from as being a memory location. Therefore, if the microprocessor is sending a signal to the outside world through an interface adapter, it would consider that interface adapter to be a memory location. It cannot distinguish between a memory location and any other location where signals go to or come from.

In the simple program discussed in Chapter 2, the final answer can be delivered from the accumulator to the outside world by using a memory location which is stored by the program.

If the microprocessor is implementing a program, you can use a logic probe on each of the address pins and you should see pulses on those pins.

Pins 26 through 32 are marked with the letter D. The D stands for data. The microprocessor is involved with two different kinds of signals. One is the address and the other is the data which tells what the actual signal is.

The difference between data and address is confusing to readers studying the microprocessor for the first time. However, the simple program that was discussed in Chapter 2 should make the distinction between these two signals very clear.

Pin 34 is marked R/$\overline{\text{W}}$. This pin is used for operating certain types of memory. It tells the memory whether the microprocessor is writing information (data) into memory or if it is reading that information (data) out of the memory. In this sense the memory is very much like your brain. If you want to put data in, that is one thing. If you want to take data out, that is another thing.

Most of us can't put data into and take data out of, our brains at the same time. Some people appear to be able to do this but it might be that they are time sharing the signals very much like most dual trace oscilloscope time shares two traces.

If you think you can do both, make a column of ten numbers of one million and greater. Add that column and at the same time talk into a tape recorder and list the telephone numbers of your friends and where you work. What you will find yourself doing is adding, then talking, then adding and then talking. This is the way the microprocessor really works and it controls the two functions depending upon whether there is a one or zero on the read/write pin.

Pin 35 has no connection. Note also that **Pin 38** has no connection.

Pin 36 is marked DBE which stands for *data bus enable*. That pin saves a lot of burned up boards. It simply prevents the information from coming in and going out at the same time. You will find the Phase II clock signal on pin 36.

Pin 39 is identified as TSC. That stands for *three state con-*

trol. A three state control enables the microprocessor to make its address pins open circuited when it is not delivering data. Therefore, outboard integrated circuits can't be addressed when the three-state control is in a logic 0 condition.

Pin 40 is marked reset. When you first turn the microprocessor on, the reset pin is low. After a minimum of eight cycles the reset pin is allowed to go high and the reset cycle begins.

The actual program for restarting the microprocessor from an OFF condition is located in the ROM.

All of the terms that have been discussed with reference to the 6800 are Motorola terminology. If you are dealing with a different manufacturer, then some of the abbreviations of the pins will be different. However, it is not difficult to adjust to these differences.

As an example, in a Motorola system, the letters CS are used for *chip select.* This input to an integrated circuit is required in order for it to receive or deliver a signal. Another manufacturer calls the same type of terminal DS which means *device select.*

Motorola uses MMI and IRQ to designate interrupt pins. Another manufacturer uses INT.

If you are dealing with a wide variety of microprocessors it would be to your advantage to photocopy some forms like the one shown in Fig. 3-3. Each time you encounter a new microprocessor, you can take the time to fill in the form with the various pin designations and at the same time record the signals for each pin when the microprocessor is operating properly.

Of course, there will be some changes in those pin signals depending upon what type of job the microprocessor is doing. But, you can very quickly assemble some good troubleshooting guides by making your own notes.

THE 6802 MICROPROCESSOR

This microprocessor is very similar to the 6800. One major exception is that it has an internal clock signal.

The 6802 pin-out diagram is shown in Fig. 3-4. The pins that are different from the 6800 are described here.

Pin 3 is marked MR (which stands for memory ready). This is a supplement for the ENABLE pin and gives the microprocessor more versatility. Specifically, it permits the microprocessor to interface with slower memories. If this pin is not used it should be tied (electrically) to VCC on pin 8.

Pin 35 is for VCC standby. This pin supplies the dc voltage to the first 32 bytes of RAM as well as the RAM ENABLE (RE)

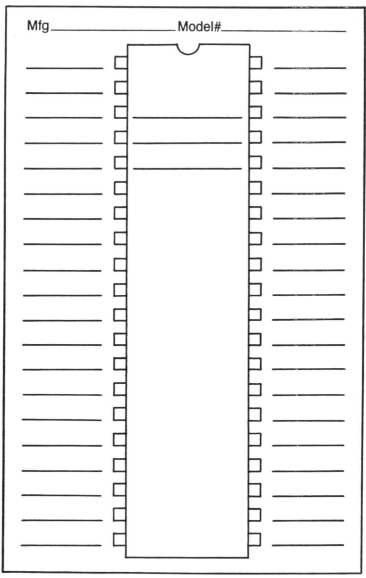

Mfg_____ Model#_____

Fig. 3-3. When you are working with microprocessor equipment it is a good idea to use this type of aid for troubleshooting.

control logic. Retention of RAM data is guaranteed during power up, power down or standby by virtue of the fact that RAM voltage is delivered on this pin. (This applies only to the first 32 bytes of RAM.)

Pin 36 is the RAM ENABLE (RE) input. This controls the on-board RAM built into the 6802 microprocessor. When this pin is in the low state, the on-board RAM is disabled. When the pin is in the high state, the on-board RAM is enabled.

Pin 37 is marked E for ENABLE. This pin supplies the clock for the microprocessor and the rest of the system. It is a single phase TTL compatible clock.

Pins 38 and 39 are for the crystal clock input. An AT cut crystal must be connected to these pins to provide the frequency standard for the internal clock signal.

Dedicated Microprocessors

A dedicated microprocessor cannot be programmed in the field. It is programmed at the factory and it is used for only one application. As an example, the microprocessor in the tuner of a television set is dedicated. In order that they don't give the whole ball game away, some of the manufacturers consider the inputs and outputs to such microprocessors as being proprietary. That's a fancy word that means they are not going to tell you what the signals are at the various pins. When that happens, it becomes essential

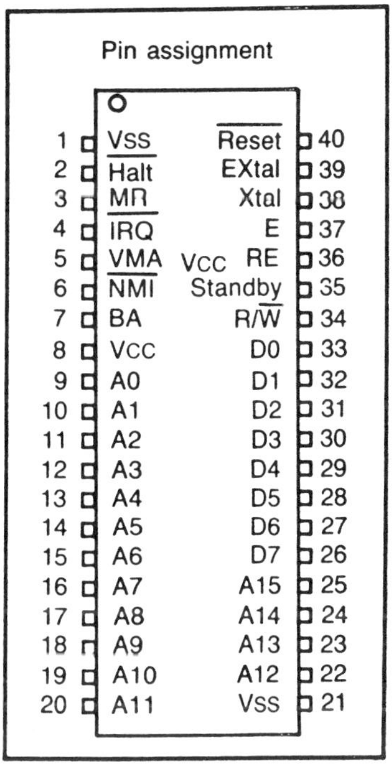

Pin assignment

1	VSS	Reset	40
2	Halt	EXtal	39
3	MR	Xtal	38
4	IRQ	E	37
5	VMA	VCC RE	36
6	NMI	Standby	35
7	BA	R/W	34
8	VCC	D0	33
9	A0	D1	32
10	A1	D2	31
11	A2	D3	30
12	A3	D4	29
13	A4	D5	28
14	A5	D6	27
15	A6	D7	26
16	A7	A15	25
17	A8	A14	24
18	A9	A13	23
19	A10	A12	22
20	A11	VSS	21

Fig. 3-4. This 6802 microprocessor is very similar to the 6800.

for you to make your own records and utilize forms like the one shown in Fig. 3-3.

As an additional note, dedicated microprocessors very often have more pins. That makes it possible to have the memories internal to the microprocessor rather than using a separate memory integrated circuit.

You may also find phase locked loops inside of these microprocessors and other specialized circuits that reduce the number of connections on the microprocessor board.

You have to be very careful when you take these large microprocessors out of a circuit. Don't try to pry them from one end. A 60- pin microprocessor will very likely crack in the middle if you start prying from one end.

Be very careful when you are probing with a logic probe that you don't let it slip between pins and thus produce a short circuit. If the microprocessor wasn't bad before you started probing, that will make a microprocessor replacement necessary.

When the microprocessor is surface mounted you have to unsolder the pins—one at a time—and bend them back. When you have them all unsoldered you have to twist the microprocessor to break the epoxy tack between it and the circuit board. (This applies to all surface-mounted components.)

You don't need to use epoxy when you replace a surface mounted component. It is simply a manufacturing technique to hold the components while they are being soldered.

Use tweezers to handle surface mounted components. Broken corners, scratches, etc. *can change their value*, so don't try to re-use damaged components. Also, handle new components with care when you are replacing damaged ones.

PROGRAMMED REVIEW

(See the instructions for the Programmed Review in Chapter 1 before using this section.)

Block 1

A microprocessor that can only do one type of work, and cannot be reprogrammed, is called a

(a) Class I microprocessor. *Go to Block 8.*
(b) dedicated microprocessor. *Go to Block 12.*

Block 2

The correct answer to the question in Block 23 is (a). Microprocessors are identified by the number of data bits. So, the 6802 is an 8-bit microprocessor.

Here is your next question: Which of the following pins has the ability to put the address lines in a high-impedance condition?

(a) R/$\overline{\text{W}}$. *Go to Block 19.*
(b) TSC. *Go to Block 13.*

Block 3

The correct answer to the question in Block 6 is (a). When the microprocessor is no longer working in a program and its address lines are available, it is said to be in a wait (WAI) condition. The microprocessor sends a signal on the BA pin at that time. The signal is also sent when the microprocessor is in a HALT condition.

Here is your next question: A certain pin on a memory IC is labeled R/$\overline{\text{W}}$. In order to write information into this memory, there must be a

(a) logic 1 on this terminal. *Go to Block 20.*
(b) logic 0 on this terminal. *Go to Block 14.*

Block 4

Your answer to the question in Block 13 is not correct. Read the question again, then *go to Block 22*.

Block 5

Your answer to the question in Block 15 is not correct. Read the question again, then *go to Block 11*.

Block 6

The correct answer to the question in Block 9 is (a). You should also look for a clock signal on any other pin where the clock signal is used. For example, there is a clock signal on DBE (pin 36) of the 6802.

Here is your next question: On the microprocessor the terminal marked BA means

(a) bus available. *Go to Block 3.*

(b) branch always. *Go to Block 24.*

Block 7

Your answer to the question in Block 23 is not correct. Read the question again, then *go to Block 2.*

Block 8

Your answer to the question in Block 1 is not correct. Read the question again, then *go to Block 12.*

Block 9

The correct answer to the question in Block 22 is (b). One of the bits in the microprocessor's condition code register determines whether or not an interrupt request on this terminal will be serviced. In other words, there is a bit on the condition code register that can mask an interrupt on this pin. If the interrupt is not masked, then the microprocessor will finish the job it is doing at the time the interrupt arrives. Immediately after finishing the task it is working on, it will store all of the numbers that are in the index register, the programmed counter, the accumulators and the condition code registers. The numbers are stored in the section of memory referred to as the *stack.*

After the interrupt has been serviced, these numbers are taken out of the stack and put back in the proper registers. Then the microprocessor continues with the program it was working on when the interrupt arrived.

Here is your next question: If you are troubleshooting a microprocessor the first step is to make sure that it has the required power supply voltages. Then, look for signals on

(a) the clock terminal(s). *Go to Block 6.*
(b) the $\overline{\text{IRQ}}$ terminal. *Go to Block 18.*

Block 10

Your answer to the question in Block 14 is not correct. Read the question again, then *go to Block 26.*

Block 11

The correct answer to the question in Block 15 is (b). This number is often referred to as 64K.

Here is your next question: If the microprocessor is working properly, you may find a clock signal on the terminal marked

(a) $\overline{\text{HALT}}$. *Go to Block 16.*
(b) DBE. *Go to Block 23.*

Block 12

The correct answer to the question in Block 1 is (b). Dedicated microprocessors are used in many consumer products such as TV tuners and microwave ovens.

Here is your next question: On which of the following pins should there be zero volts in a 6802 microprocessor circuit?

(a) A pin marked VSS. *Go to Block 15.*
(b) A pin marked VGG. *Go to Block 21.*

Block 13

The correct answer to the question in Block 2 is (b).

Here is your next question: On which of the following pins should there be a $+5V$ in a 6802 microprocessor circuit?

(a) A pin marked VBB. *Go to Block 4.*
(b) A pin marked VCC. *Go to Block 22.*

Block 14

The correct answer to the question in Block 3 is (b). The overbar on the W means it is active low.

Here is your next question: In order to get the microprocessor ready for action after the power is first turned ON, it is necessary (after a minimum of eight cycles) to deliver a logic 0 to the pin marked

(a) $\overline{\text{RESET}}$. *Go to Block 26.*
(b) VCC. *Go to Block 10.*
(c) Neither answer is correct. *Go to Block 25.*

Block 15

The correct answer to the question in Block 12 is (a). There is no pin marked "VGG" on the microprocessor.

Here is your next question: If a microprocessor has 16 address lines, the number of addresses it can select is

(a) 256. *Go to Block 5.*
(b) 65,536. *Go to Block 11.*

Block 16

Your answer to the question in Block 11 is not correct. Read the question again, then *go to Block 23.*

Block 17

Your answer to the question in Block 22 is not correct. Read the question again, then *go to Block 9.*

Block 18

Your answer to the question in Block 9 is not correct. Read the question again, then *go to Block 6.*

Block 19

Your answer to the question in Block 2 is not correct. Read the question again, then *go to Block 13.*

Block 20

Your answer to the question in Block 3 is not correct. Read the question again, then *go to Block 14.*

Block 21

Your answer to the question in Block 12 is not correct. Read the question again, then *go to Block 15.*

Block 22

The correct answer to the question in Block 13 is (b). There is no terminal marked VBB.

Here is your next question: When a pin is marked $\overline{IRQ}$, it means that in order to request an interrupt the logic level on that pin must be

(a) logic 1. *Go to Block 17.*
(b) logic 0. *Go to Block 9.*

Block 23

The correct answer to the question in Block 11 is (b). When you are troubleshooting you should check for clock signals at all pins where they are required.

Here is your next question: The number of data lines on the 6802 microprocessor is

(a) 8. *Go to Block 2.*
(b) 16. *Go to Block 7.*

Block 24

Your answer to the question in Block 6 is not correct. Read the question again, then *go to Block 3.*

Block 25

Your answer to the question in Block 14 is not correct. Read the question again, then *go to Block 26.*

Block 26

The correct answer to the question in Block 14 is (a). You have now completed the programmed review.

EXPERIMENTS

In the experiments for this chapter you construct some of the basic circuits used in a microprocessor.

Registers

Registers are used for temporarily storing information in the microprocessor. As an example, an accumulator—which is a type of register—stores a number that is to be used by the arithmetic logic unit.

The 8 digit numbers stored in an 8-bit microprocessor are called *words*. (In a 4-bit microprocessor a word is only 4 bits long.) Also, eight bits is referred to as a *byte*. (Four bits is called a *nibble*, and sixteen bits is called two bytes.)

Registers are also used in peripheral equipment such as the peripheral interface adapter (PIA) and the asynchronous communications interface adapter (ACIA).

Because of their importance in microprocessor systems, you

will encounter the subject of registers throughout this book.

The four basic types of registers are illustrated in Fig. 3-5. All of these types are used in microprocessor systems. The registers discussed in relation to the program of Chapter 2 are PIPO types. Examples of registers are the 7495, 74164 and 74165.

The simplified version of one register bit in Fig. 3-6 demonstrates the steps that are taken in storing a word.

A switch is used to select the bit to be stored. In the microprocessor the bits to be stored come from memory, or some internal source such as the ALU output.

A switch is used to operate the enable so that the bit to be stored can be delivered to the data flip-flop. In practice, a logic 1 would be delivered from the control bus to operate this enable.

A clear switch is used to erase the stored bit. In most registers

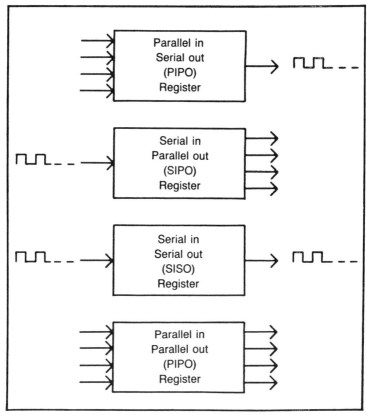

Fig. 3-5. The four types of registers are shown here. The serial in/serial out type is not often used in microprocessor systems.

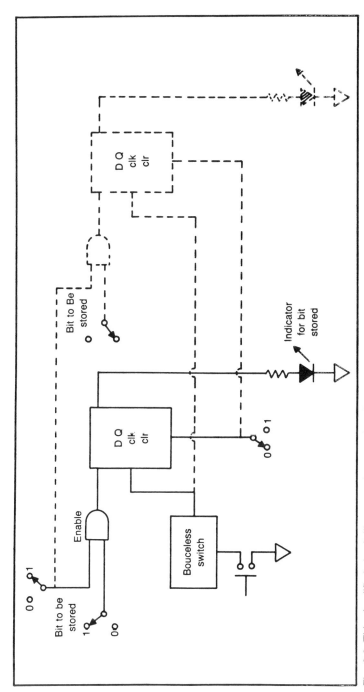

Fig. 3-6. This simplified circuit illustrates the procedure for loading and retrieving a number in a register.

this is not necessary because the bit to be stored will write over the one that was previously stored.

The bounceless switch serves as the clock pulse. The data cannot be entered into the D flip-flop until a clock transition takes place.

The LED shows the type of bit (1 or 0) that has been stored.

Build the one-bit register of Fig. 3-6 and show that it can be used to store either a logic 1 or logic 0.

Enlarge the register so that four bits can be stored. Note that they are all entered at the same time, so you need only one enable switch and one bounceless switch. Also, one clear switch is needed. Store number 7 (0111) and then overwrite it with the number 8 (1000). Note that the LEDs always show the last number that has been stored.

The four-bit register that you have built can be used as a register in a microprocessor. It can also be used for storing one byte in a larger memory.

The 7495 is an integrated circuit register for parallel in and parallel out. You should obtain one of these registers and experiment with it to get a better understanding of how information is put into registers and taken out of registers.

SELF TEST

(Answers at the end of the chapter)

1. Microprocessors are identified by the number of _____ bits they can work with, or, the number of _____ terminals they have.

2. You expect to see clock signals on the ϕ, and ϕ_2 clock terminals. Name an additional terminal on the 6802 where you should see a clock signal during normal operation.

3. When the operation of the microprocessor is to be stopped a logic 0 is delivered to the terminal called _____.

4. Which of the following pins will be put in the high-impedance state with the proper signal to the TSC terminal?

   (a) Data pins
   (b) Address pins

5. To begin microprocessor operation after it has been shut down, a signal is delivered (after a minimum of 8 cycles) to the _____ terminal.

6. What value of power supply voltage is required for operating the 6802?

7. Which type of interrupt can be prevented (masked) by a bit in the condition code register?

8. The microprocessor condition code register cannot mask an interrupt signal that arrives at pin number _____.

9. To write information into memory, a (*logic 1 or logic 0*) must be delivered to its R/$\overline{\text{W}}$ terminal.

10. An eight-bit microprocessor has eight
    (a) data lines.
    (b) address lines.

Answers to the Self Test

1. data - data
2. <u>Pin 12</u>
3. $\overline{\text{HALT}}$
4. (b)
5. reset
6. <u>+5V</u>
7. $\overline{\text{IRQ}}$
8. 6
9. logic 0
10. (a)

Chapter 4

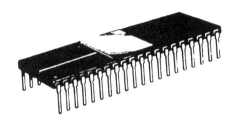

The ALU

In this chapter—

- The theory of bit slicing
- Operation of a typical ALU
- Advantage of Schottky ALU
- Pin identification of typical ALU
- Experiments with the ALU

You studied the operation of the arithmetic logic unit in Chapter 2 when a program was followed through a microprocessor. As its name implies, this unit is capable of doing simple arithmetic and logic operations. It is made entirely with logic gates which are put into circuits such as half adders, full adders, exclusive ORs, etc.

Figure 4-1 shows two ways that arithmetic logic units are used in the two most popular microprocessors. In the 8080 the numbers that are to be introduced to the ALU are held in an accumulator and a register. The 6800 uses two accumulators.

In this chapter we will not deal with the combinations of gates that are used to make the arithmetic and logic functions. Instead, we will deal directly with an example of an ALU that can be breadboarded for experimentation.

BIT SLICING

Bit slicing is the reason for these integrated circuit ALUs be-

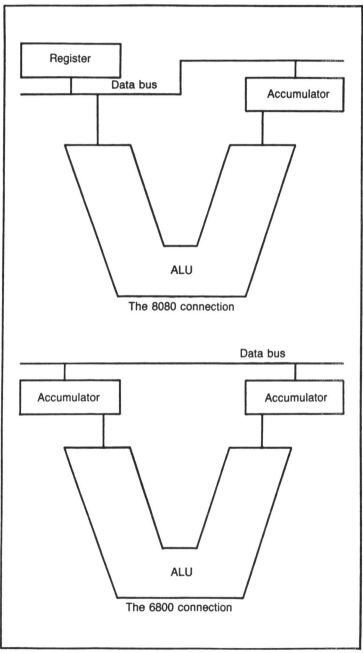

Fig. 4-1. This illustration shows two different ways of holding numbers for the ALU. In both cases temporary memories are used.

ing available. A bit slice is a microprocessor that is put together with individual integrated circuits. An example of a bit slice is shown in Fig. 4-2. There are some obvious disadvantages in making a microprocessor this way.

One is that you do not have a stored program with op codes and micro instructions available. The person who puts the bit slice together must make the required internal and external programs. There is also a somewhat greater area required for putting the bit slice together compared to having everything in a single integrated circuit package.

Last, but not least, putting a bit slice together requires some very tedious design work so that all of the signals arrive at the required places at exactly the right time. All of these disadvantages would seem to make the bit slice an undesirable technology, but there is one advantage of bit slicing over integrated circuit microprocessors.

The bit slice is much faster in its operation. For that reason you will see bit slices used whenever speed is the criteria for design.

THE 74181 ALU

In some integrated circuits designed for bit slicing the manufacturer includes registers and timing circuitry. The 74181, however, does not have those features. This makes it very useful for experimenting with the ALU.

The pin-out diagram of the 74181 is shown in Fig. 4-3. The input data to this 24-pin integrated circuit is on pin 1 and 2 and also 18 through 23. The terminals marked 'A' combine to serve one input to the ALU and those marked 'B' combine to serve the other input. In practice the inputs to 'A' and 'B' might come from two accumulators.

This is a 4-bit ALU, but that does not limit its application to 4-bit data lines. The ALUs can be operated in parallel to accommodate any desired number of bits.

Pins 3 through 6 are designated with the letter S. These pins are used to encode the ALU so that it will perform any of the functions listed in Table 4-1.

Refer to Table 4-1, for example, and note that in the *logic mode* the inputs at A are inverted when S0 through S3 are in a low condition. We will return to this table later in discussing the use of the 74181 in the Experiment section.

There are two possible modes for using this ALU. These modes are determined by the input at M (pin 8). When M is in a high con-

Table 4-1. Mode Select-Function Table.

Mode Select Inputs				Active High Inputs & Outputs	
S_3	S_2	S_1	S_0	Logic (M = H)	Arithmetic** (M = L) (C_n = H)
L	L	L	L	$\overline{A}$	A
L	L	L	H	$\overline{A + B}$	A + B
L	L	H	L	$\overline{A}B$	A + $\overline{B}$
L	L	H	H	Logical 0	minus 1
L	H	L	L	$\overline{AB}$	A plus A$\overline{B}$
L	H	L	H	$\overline{B}$	(A + B) plus A$\overline{B}$
L	H	H	L	A $\oplus$ B	A minus B minus 1
L	H	H	H	A$\overline{B}$	AB minus 1
H	L	L	L	$\overline{A}$ + B	A plus AB
H	L	L	H	A $\oplus$ B	A plus B
H	L	H	L	B	(A + $\overline{B}$) plus AB
H	L	H	H	AB	AB minus 1
H	H	L	L	Logical 1	A plus A*
H	H	L	H	A + $\overline{B}$	(A + B) plus A
H	H	H	L	A + B	(A + $\overline{B}$) plus A
H	H	H	H	A	A minus 1

Mode Select Inputs				Active Low Inputs & Outputs	
S_3	S_2	S_1	S_0	Logic (M = H)	Arithmetic** (M = L) (C_n = L)
L	L	L	L	$\overline{A}$	A minus 1
L	L	L	H	$\overline{AB}$	AB minus 1
L	L	H	L	$\overline{A}$ + B	A$\overline{B}$ minus 1
L	L	H	H	Logical 1	minus 1
L	H	L	L	$\overline{A + B}$	A plus (A + $\overline{B}$)
L	H	L	H	$\overline{B}$	AB plus (A + $\overline{B}$)
L	H	H	L	$\overline{A \oplus B}$	A minus B minus 1
L	H	H	H	A + $\overline{B}$	A + $\overline{B}$
H	L	L	L	$\overline{A}B$	A plus (A + B)
H	L	L	H	A $\oplus$ B	A plus B
H	L	H	L	B	A$\overline{B}$ plus (A + B)
H	L	H	H	A + B	A + B
H	H	L	L	Logical 0	A plus A*
H	H	L	H	A$\overline{B}$	AB plus A
H	H	H	L	AB	A$\overline{B}$ plus A
H	H	H	H	A	A

L = LOW voltage
H = HIGH voltage level
  *Each bit is shifted to the next more significant position.
**Arithmetic operations expressed in 2s complement notation.

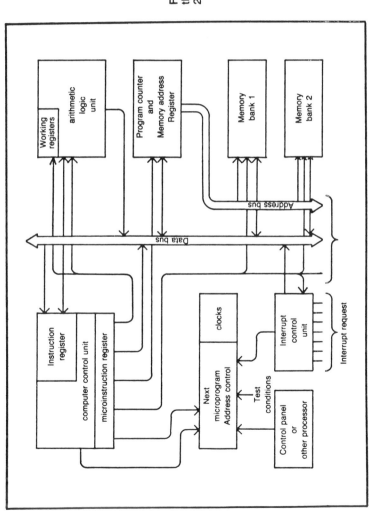

Fig. 4-2. Compare this bit slice with the microprocessor shown in Fig. 2-1.

73

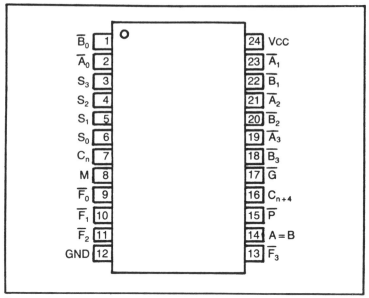

Fig. 4-3. The pinout for the 74181 ALU is shown here.

dition (logic 1) the ALU performs logic operations. When that pin is in a low condition (logic 0) the ALU performs arithmetic operations.

Pins 15, 16 and 17 are used for carry operations when ALUs are connected in parallel. The signal outputs from the ALU are on pins 9, 10, 11 and 13.

The supply voltage (Vcc) for the ALU must be maintained between 4.5 and 5.5 volts. The nominal value is +5 volts. The common connection for the integrated circuit is at pin 12.

There is a Schottky version of a 74181. It is identified by the letters 74S181. The Schottky version has a propagation delay of only 11 nanoseconds compared to a propagation delay of 22 nanoseconds for other versions of the ALU. The power supply current requirement is 120 milliamps for the Schottky version. There is a low power Schottky (74LS181) which requires only 21 milliamps. The regular 74181 requires 91 milliamps of current for the power supply.

Pin 14 of the 74181 is labeled A = B. This is an open collector terminal. In other words, it cannot be operated unless a pull up resistor is connected between this pin and the power supply.

The output at pin 14 can be used to indicate equivalence when the unit is in the subtract mode. If all the outputs (at the F terminal) are high the output at A = B will also be high.

The versatile 74181 ALU can be operated with either active high inputs and outputs or active low inputs and outputs—depending upon whether pin 7 (C_N) is in a high or low condition. When C_N has a logic 1 on its terminal, the inputs and outputs are active high. When C_N is switched to a low condition the inputs and outputs of the ALU are active low.

PROGRAMMED REVIEW

(See the instructions for the Programmed Review in Chapter 1 for using this section.)

Block 1

The pins designated by the letter 'S' on the 74181 ALU are used for

(a) single stepping. *Go to Block 8.*
(b) encoding. *Go to Block 12.*

Block 2

The correct answer to the question in Block 23 is (b). Here is your next question: Which of the following is the more expected application of an ALU?

(a) Bit slice. *Go to Block 10.*
(b) Cash register operation. *Go to Block 15.*

Block 3

Your answer to the question in Block 22 is not correct. Read the question again, then *go to Block 9.*

Block 4

The correct answer to the question in Block 25 is (a). The pull up resistor is like a load resistor. Without it there is no dc delivered to the active (transistor) component.

Here is your next question: An ALU is made with

(a) logic gates. *Go to Block 22.*
(b) operational amplifiers. *Go to Block 18.*

Block 5

The correct answer to the question in Block 20 is (b). In most cases it is not necessary to clear a register before inserting a new number. You simply write over the number that is already there.

Here is your next question: An 8-bit ALU can be made by connecting

(a) two 4-bit ALUs in parallel. *Go to Block 13.*
(b) two 4-bit ALUs in series. *Go to Block 17.*

Block 6

The correct answer to the question in Block 12 is (a). A bit slice is much faster than a microprocessor—especially when emitter coupled logic (ECL) is used in the design. ECL is the fastest logic family.

Here is your next question: The Schottky version of the ALU has

(a) a longer propagation delay. *Go to Block 16.*
(b) a shorter propagation delay. *Go to Block 20.*

Block 7

Your answer to the question in Block 25 is not correct. Read the question again, then *go to Block 4.*

Block 8

Your answer to the question in Block 1 is not correct. Read the question again, then *go to Block 12.*

Block 9

The correct answer to the question in Block 22 is (c).

Here is your next question: Two accumulators are used to store numbers for the ALU in the

(a) 8080. *Go to Block 14.*
(b) 6800. *Go to Block 23.*

Block 10

The correct answer to the question in Block 2 is (a). Modern cash registers use dedicated microprocessors or computers.

Here is your next question: The supply voltage for the 74181 ALU is

(a) ± 12V unregulated. *Go to Block 19.*
(b) $+5$V regulated. *Go to Block 24.*
(c) ± 12V regulated. *Go to Block 28.*

Block 11

Your answer to the question in Block 12 is not correct. Read the question again, then *go to Block 6.*

Block 12

The correct answer to the question in Block 1 is (b). There is no such thing as "single stepping" in an ALU.

Here is your next question: Which of the following is a disadvantage of using a bit slice design compared with using an integrated circuit microprocessor?

(a) Manufacturer's op code not available. *Go to Block 6.*
(b) Slow speed. *Go to Block 11.*

Block 13

The correct answer to the question in Block 5 is (a).

Here is your next question: A microprocessor that is put together with individual integrated circuits is called a

(a) chip package. *Go to Block 21.*
(b) bit slice. *Go to Block 25.*

Block 14

Your answer to the question in block 9 is not correct. Read the question again, then *go to Block 23.*

Block 15

Your answer to the question in Block 2 is not correct. Read the question again, then *go to Block 10.*

Block 16

Your answer to the question in Block 6 is not correct. Read

the question again, then *go to Block 20.*

Block 17

Your answer to the question in Block 5 is not correct. Read the question again, then *go to Block 13.*

Block 18

Your answer to the question in Block 4 is not correct. Read the question again, then *go to Block 22.*

Block 19

Your answer to the question in Block 10 is not correct. Read the question again, then *go to Block 24.*

Block 20

The correct answer to the question in Block 6 is (b). Whenever you see Schottky in the name of a logic IC you know it is faster than a comparable IC that does not employ a Schottky configuration.

Here is your next question: The first step in using the ALU in the 6800 is to

(a) clear the register. *Go to Block 29.*
(b) load the accumulators. *Go to Block 5.*

Block 21

Your answer to the question in Block 13 is not correct. Read the question again, then *go to Block 25.*

Block 22

The correct answer to the question in Block 4 is (a). Given enough time and enough logic gates you could make your own ALU or microprocessor. In fact, most of the integrated circuits used in a microprocessor system are made with logic gates.

Here is your next question: The inputs and outputs of the 74181 ALU

(a) are active low. *Go to Block 26.*
(b) are active high. *Go to Block 3.*

(c) can be either active low or active high. *Go to Block 9.*

Block 23

The correct answer to the question in Block 9 is (b). Accumulators could be used to store the A and B inputs for the 74181 ALU.

The 8080 uses an accumulator and a register to store numbers for the ALU input and output numbers.

Here is your next question: The 74181 ALU is either in a logic or an arithmetic mode depending upon whether there is a 0 or 1 delivered to the pin marked

(a) L/A. *Go to Block 27.*
(b) M. *Go to Block 2.*

Block 24

The correct answer to the question in block 10 is (b).

Here is your next question: If the inputs to the A and B terminals of the 74181 ALU are all 1's, and the ALU is wired as an AND, should all of the outputs be a logic 1 or a logic 0? *Go to Block 30.*

Block 25

The correct answer to the question in Block 13 is (b). Bit slice integrated circuits are readily available. In many bit slice packages the timing and interconnection problems are solved by the manufacturer. So, instead of an ALU integrated circuit, there is an IC package that includes other logic-bussed units.

Here is your next question: An integrated circuit with an open-collector terminal requires

(a) a pull up resistor. *Go to Block 4.*
(b) a pull down resistor. *Go to Block 7.*

Block 26

Your answer to the question in Block 22 is not correct. Read the question again, then *go to Block 9.*

Block 27

Your answer to the question in Block 23 is not correct. Read the question again, then *go to Block 2.*

79

Block 28

Your answer to the question in Block 10 is not correct. Read the question again, then *go to Block 24.*

Block 29

Your answer to the question in Block 20 is not correct. Read the question again, then *go to Block 5.*

Block 30

The output pins should all have a logic 1 level because 1 AND 1 = 1. You have now completed the Programmed Review.

EXPERIMENTS

The experiments in this chapter will give you practice in working with the ALU.

Using the 74181 ALU to Add Two Numbers

Instead of using registers, use switches in these experiments to hold the numbers that are being delivered to the A and B sections of the ALU. These switches must be able to switch between logic 1 and logic 0.

Do not assume that an open line into the ALU will represent a logic 0. That is never good practice in logic circuitry. Use the switch arrangement shown in Fig. 4-4.

The outputs from the ALU ($\overline{F0}$ through $\overline{F3}$) should be wired to LEDs through the appropriate current limiting resistors. A value of 330 or 470 ohms is recommended for this resistor. A high resistance results in a very low output current from the device.

Set the input switches to add the numbers 3 (0011) and 4 (0101). The LEDs at the output should display the number 7. *Remember that you must set the mode switch in order to get this ALU into an addition mode, and you must set the input terminals for coding (S0 through S3) to get it into the specific arithmetic mode that you have chosen.* (In this case you will use A + B.)

Using the 74181 ALU in the Logic Mode

In this experiment you will set up the ALU to work as a burglar alarm. To do this, you will set all of the A inputs to a logic 1. These inputs could be hard wired, but if you still have the

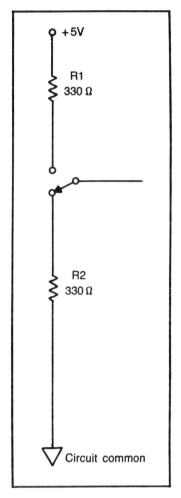

Fig. 4-4. This simple circuit is used for switching between logic 1 and logic 0. In many cases the pull-down resistor (R2) is eliminated so that the switch goes directly to logic 0.

switches wired in from the previous experiment, just simply set them all to a logic 1.

The inputs to be delivered to the B inputs represent the inputs from switches on windows and/or doors. When the windows are closed you will presume that a logic 1 is present at these inputs.

When you have set your mode input and your S terminals to perform an AND operation, then all of the outputs should also be logic 1. In other words, all of the LEDs will be lit. However, if one of the windows is opened (which is represented by changing AB input from a 1 to a 0) the LED corresponding to that window will not be lit.

So, this burglar alarm not only permits you to determine that a window is open, but it also permits you to determine which window is open.

Instead of using LEDs, an alarm could be used, or an alarm could be used in conjunction with the LEDs.

If you are having trouble with these experiments, make sure that you have set the proper mode and also that you are using the active low operation so that the LEDs are normally ON.

SELF TEST

(Answers at the end of the chapter)

1. A microprocessor made with individual integrated circuits is called a _____.

2. Compared to a single integrated circuit microprocessor (like the 6802), a bit slice is

   (a) slower.
   (b) faster.

3. Compared to a single integrated circuit microprocessor (like the 6802), a bit slice is

   (a) easier to program.
   (b) more difficult to program.

4. What is the name of the pin used to determine whether the ALU is performing arithmetic or logic?

5. The 74181 works with a power supply voltage of _____.

6. To get the ALU set for logic operation it is necessary for pin 8 to be

   (a) LOW.
   (b) HIGH.

7. The pin that determines whether the ALU is in an active high or active low condition is marked _____.

8. Assume the ALU is in the active high mode of operation. What is the S code for A AND B?

9. The terminal marked $\overline{G}$ is used for _____.

10. There are three versions of the 74181 ALU. They are 74181, 74LS181, and 74S181. Which of these is fastest?

Answers to the Self Test

1. bit slice
2. (b)
3. (b)
4. M (for Mode)
5. +5 V
6. (b)
7. C_N
8. S0 = H; S1 = H; S2 = L; S3 = H
9. generating a carry to the next ALU (if used).
10. 74S181. You pay for this fastest operation with a higher power supply current requirement.

Chapter 5

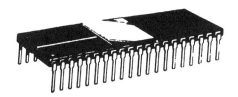

The ROM

In this chapter—

- Static versus dynamic memories
- Volatile and non-volatile memories
- Comparison of RAM and ROM
- Examples of serial and parallel memory
- The ideas behind CCD and bubble memories

Memory is the key to the operation of all microprocessor and computer systems. You will remember that the microprocessor was originally designed to implement memories, and it was designed at the request of a memory manufacturer.

In this chapter we will look at the two basic kinds of memory: the RAM and the ROM. The emphasis will be on the ROM. In the next chapter the RAM will be studied in greater detail.

SOME BASIC TERMS

All memories can be divided into two main categories, *dynamic* and *static*. A dynamic memory is one that must be continually refreshed. Basically, the dynamic memory relies upon the charge of a capacitor to store a logic 1. In other words, when the capacitor is charged, a logic 1 is stored. When the capacitor is discharged, a logic 0 is stored.

Since capacitors cannot maintain their charge over a long period of time, it is necessary to recharge (or refresh) this type of memory.

Refreshing memories is not as much of a problem as it used to be. Today there are integrated circuits specifically designed for that purpose. Still, it is not as convenient as using a static memory which does not have to be refreshed.

Considering the inconvenience of recharging the capacitor (dynamic) memories, it would seem that the dynamic memory would not have much application. However, it has one characteristic that makes it essential for certain microprocessor and computer systems: the dynamic memory is very, very fast.

A second important characteristic is that dynamic memories have a very high density, so a very large amount of memory can be put onto a relatively small amount of "real estate."

With a static memory, it is not necessary to have a refresh circuit. Instead, a logic level is usually stored by setting a simple flip-flop in a high or low condition. The flip-flops are put in the high condition to store a logic 1, and they are put in a low condition for storing a logic 0.

As you know, once a flip-flop is put into a high or low condition, it stays that way until an unsettling input pulse arrives to change its condition. That means the information in the static memory can remain for long periods of time without the need for refreshing.

Another way of categorizing memories is by whether or not they are *volatile*. A *volatile memory* is one that loses all of the stored information when the power supply is turned off. Both dynamic and static memories can be volatile.

A *non-volatile memory* maintains the information stored even when it is completely out of the circuit.

A third category is *volatile protected*. This is a volatile memory which is connected to a power supply that has a stand-by operation. If the main power is lost, the stand-by power kicks in and maintains the voltage supplied to the memory. Stand-by power is battery operated and it is used as a back-up for power supplies that obtain their power from the power line.

As a general rule, dynamic memories are volatile because voltage is needed for refreshing. Static memories are also volatile because they require a voltage to hold their stored information.

Read Only Memories, or ROMs (to be discussed) are non-volatile.

A very important method of categorizing memories is according to whether information can be stored and retrieved very quickly as in the case of a RAM, or is stored one time and cannot be changed as in the case of a ROM. Figure 5-1 illustrates the general ideas behind some memories. Figure 5-1(A) compares the idea of ROM to that of RAM.

The ROM is made with a fusible link. When this device is programmed, the fuse is burned open for one logic level, or left intact for another logic level. This is accomplished by a so-called *ROM burner*. Once the fusible link has been burned open, it cannot be changed.

In the case of the RAM, which is represented by a switch, it can be open for one logic level and closed for the other, and it can be readily changed when the information is to be changed. In prac-

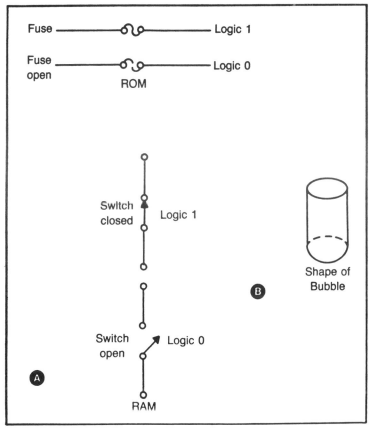

Fig. 5-1. (A) The ROM represents a permanent logic 1 or logic 0. The RAM can be changed back and forth for each logic level. (B) Bubble memory.

tice, the switch would be replaced with a flip-flop, but the principle is still the same. The switch shows that the bit stored can be changed.

There is a certain amount of confusion between the ROM and RAM terminology. The term RAM stands for Random Access Memory and it seems to imply that it is the only type of memory in which you can go to a certain location and retrieve information stored, or in which you can go to a certain location and store information. Actually, the ROM can also retrieve information from any stored position. *In that sense,* a ROM is a random access memory.

To get around this confusion, there was a move to change the name RAM to *Read/Write Memory.* Unfortunately, that term never got a good foothold. Today the two types are still usually referred to as RAMs and ROMs.

TYPES OF ROMS

To further confuse the terminology, there are certain types of ROM in which the information stored can be changed. It cannot be changed as readily as in a RAM, but by utilizing the proper equipment (a ROM burner) the ROMs are electromagnetically programmable. This type of memory is called PROM. The letters mean *Programmable Read Only Memory.*

Types of PROMs

There are three main classifications of PROMs. They are: EEROM, which stands for *Electrically Erasable Read Only Memory;* EPROM, which means *Electrically Programmable Read Only Memory*; and EAROM, which stands for *Electrically Alterable Read Only Memory.* They will be discussed in alphabetical order.

The contents of the EAROM can be altered without removing the integrated circuit from its place on a printed circuit board. Putting new information into an EAROM requires a much longer period of time than retrieving stored information. So, you will sometimes hear the term *read mostly* used in conjunction with this type of read only memory.

EEROMs are non-volatile, Electrically Erasable, Read Only Memories. Like the EAROM, they can be erased while they are still intact in the printed circuit board. They are similar to EAROMs in their application, but it is easier to remove stored information. An electrical signal is used for that purpose.

EPROMs must be removed from their circuit in order to be

reprogrammed. They are erased with an ultraviolet light. Usually, EPROMs have a sticker covering a window on their back that can be easily removed so an ultraviolet light can be used to erase stored information. If this sticker is accidentally removed and the EPROM is exposed for any period of time to ultraviolet light or to a fluorescent light, the information stored will be lost.

A *programmer* is used to store new information into the EPROM after it has been erased.

SOME ADDITIONAL MEMORIES

Some memories are made in such a way that they have only serial access. In other words, the information is stored and retrieved one bit at a time. A *Serial In/Serial Out shift register* is an example of a memory made this way. You cannot access any single bit in this memory without going through some of the other memory bits which have been stored. This type of memory is useful where the information is always going to be retrieved in order.

RAM and ROM memories are usually classified as parallel access. That means all of the information at one address is entered and/or retrieved at the same time.

One of the newer kinds of memory is CCD which stands for Charge Coupled Device. This type of memory stores a logic 1 by implanting a charge under the surface of a semiconductor material. The charge is stored in a location called a *potential well.*

The information in a charge coupled device is put in and taken out in serial form, so it is not a random access memory.

CCD is a volatile type of memory and a refresh circuit is required for its use. Its advantage is its high density. It can store a large amount of information on a relatively small amount of real estate.

Another type of memory which is non-volatile is the bubble memory. This device stores a logic 1 or logic 0 according to whether there is a small magnetic cylinder at the point of interest. These magnetic domains are actually cylindrical in shape as shown in Fig. 5-1(B) but viewed from the top they look like tiny bubbles.

The bubble memory is non-volatile and has a high *density.* Information is stored and retrieved in serial form.

Magnetic Bubble memories are relatively small compared to other serial devices like the charge coupled device. Being non-volatile, the magnetic bubble memory does not require a refresh circuit.

MASS MEMORIES

In addition to the memory that we have talked about here, there are mass memories such as floppy disks, magnetic tapes, optical disks, and so forth. These mass memories are mostly used in complete computer systems and they will not be discussed in this microprocessor book. However, you should understand that the ultimate goal is still the same: to store large amounts of information and be able to retrieve that information as it is needed.

THE PLA

A programmable logic array (PLA) is very closely related to read only memory. These integrated circuit devices consist of fusible links and basic transistor structures to form AND/OR logic functions. They also contain inverters. By the proper programming of a logic array, it is possible to make all of the basic logic gates. It is used in a logic circuit to implement combinational logic arrays.

There are two types of programmable logic arrays. One is programmed with a fusible link and the other is field programmable.

The advantage of any programmable logic array is that it saves a lot of space and reduces construction time of a logic circuit.

MEMORY ORGANIZATION

Figure 5-2 shows a block diagram of a typical read only memory. This one has an organization that is described as 2K × 8. It means that there are 2048 eight-bit words stored (remember that a K is equal to 1024 in computer talk), and these words can be retrieved one at a time. In other words, the output is an 8-bit word for each address input.

By contrast, a 64 by 4 memory would have 64 four-bit words stored, and they could be retrieved four bits at a time. For that memory, four bits would be one word, and the memory can be retrieved one word at a time.

The chip select (CS), chip enable (CE) and other terminology in Fig. 5-2 have already been discussed previously for other integrated circuit packages.

PROGRAMMED REVIEW

(See the instructions for the Programmed Review in Chapter 1 for using this section.)

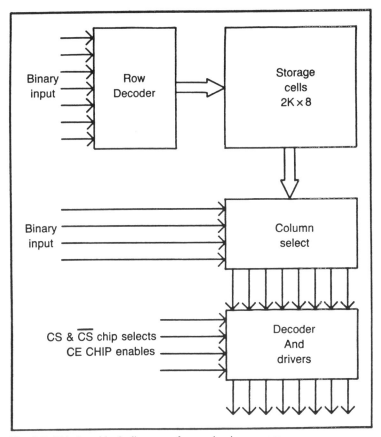

Fig. 5-2. This is a block diagram of a read-only memory.

Block 1

The original purpose of a microprocessor was to

(a) implement memory. *Go to Block 8.*
(b) control machinery. *Go to Block 12.*

Block 2

Your answer to the question in Block 5 is not correct. Read the question again, then *go to Block 20.*

Block 3

The correct answer to the question in Block 15 is (b). The advantages of speed and high density are for today's technology. In the future it is possible that static memories might overtake the dynamic RAM in those categories.

Here is your next question: Flip-flops are used to store information in some types of

(a) RAM. *Go to Block 11.*
(b) ROM. *go to Block 17.*

Block 4

The correct answer to the question in Block 9 is (b). As a general rule, RAMs are volatile and ROMs are non-volatile.

Here is your next question: "A ROM cannot be randomly accessed." This statement is

(a) correct. *Go to Block 10.*
(b) not correct. *Go to Block 18.*

Block 5

The correct answer to the question in Block 11 is (a). A ROM burner is used to open the links (as required) for storing information.

Here is your next question: Bubble memories are

(a) volatile. *Go to Block 2.*
(b) non-volatile. *Go to Block 20.*

Block 6

The correct answer to the question in Block 8 is (a). An early model television receiver used a charged capacitor as a memory to retain the channel selected. The capacitor remained charged without the need for refreshing, and it held the station for long periods without any problem.

Normally, however, capacitors in dynamic memories must be refreshed.

Here is your next question: Which of the following memories retains stored information when the power supply is turned off?

(a) RAM. *Go to Block 19.*
(b) ROM. *Go to Block 9.*

Block 7

Your answer to the question in Block 18 is not correct. Read the question again, then *go to Block 15.*

Block 8

The correct answer to the question in Block 1 is (a). Microprocessors and computers both utilize memory. Their operation can be understood by understanding how they make use of memories.

Here is your next question: In a certain memory a charged capacitor represents a logic 1 and a discharged capacitor represents a logic 0. This is an example of

(a) a dynamic memory. *Go to Block 6.*
(b) a static memory. *Go to Block 14.*

Block 9

The correct answer to the question in Block 6 is (b). Information is permanently stored in a ROM.

Here is your next question: A ROM is an example of

(a) a volatile memory. *Go to Block 16.*
(b) a non-volatile memory. *Go to Block 4.*

Block 10

Your answer to the question in Block 4 is not correct. Read the question again, then *go to Block 18.*

Block 11

The correct answer to the question in Block 3 is (a). Data flip-flops are used in RAMs. A high condition for a flip-flop represents a logic 1; and, a low condition means that a logic 0 is stored.

Here is your next question: One example of a ROM is made with a
(a) fusible link. *Go to Block 5.*
(b) saturated transistor. *Go to Block 13.*

Block 12

Your answer to the question in Block 1 is not correct. Read

the question again, then *go to Block 8.*

Block 13

Your answer to the question in Block 11 is not correct. Read the question again, then *go to Block 5.*

Block 14

Your answer to the question in Block 8 is not correct. Read the question again, then *go to Block 6.*

Block 15

The correct answer to the question in Block 18 is (b). Information in the EEROM can be electrically erased.

Here is your next question: Which of the following are advantages of dynamic memory?

(a) Convenient to use and retains memory when power supply is off. *Go to Block 21.*
(b) Very fast and can be made with high density. *Go to Block 3.*

Block 16

Your answer to the question in Block 9 is not correct. Read the question again, then *go to Block 4.*

Block 17

Your answer to the question in Block 3 is not correct. Read the question again, then *go to Block 11.*

Block 18

The correct answer to the question in Block 4 is (b). Both ROMs and RAMs can be randomly accessed. That means that information in any location in the memory can be found and used.

Here is your next question: Eventually, a fluorescent light can erase an

(a) EEROM. *Go to Block 7.*
(b) EPROM. *Go to Block 15.*

Block 19

Your answer to the question in Block 6 is not correct. Read the question again, then *go to Block 9.*

Block 20

The correct answer to the question in Block 5 is (b). The bubbles are like tiny permanent magnets that are shaped like cylinders.

Here is your next question: One kind of memory stores information in the form of charges below the surface. It is called a _____. *Go to Block 22.*

Block 21

Your answer to the question in Block 15 is not correct. Read the question again, then *go to Block 3.*

Block 22

The correct answer to the question in Block 20 is CCD (charge coupled device). You have now completed the programmed review.

EXPERIMENTS

The wide variety of ROMs used in microprocessors are programmed for each special application. Earlier in this book you studied about a monitor which takes care of some basic microprocessor operations. That monitor is actually a ROM.

A ROM may also be used to store constants such as *pi* and *epsilon.* A ROM can be used to convert a binary number into the outputs required for a seven-segment decoder. That is not to say that only ROMs are used for this purpose. In fact many decoders are made with simple logic circuitry. You cannot tell by looking at the integrated circuit whether the decoder is made from a ROM or from logic circuitry.

In this experiment we will assume that the decoder is made from a ROM. For each binary input, there is a permanent code that represents the decimal equivalent of that binary number. The decoder delivers the necessary voltages to the seven-segment display, so the decimal equivalent of the binary input will be shown.

A seven-segment decoder converts a binary number input to an output for operating a seven-segment display. In the simplified example of Fig. 5-3 the input to the decoder is 0110.

That represents the decimal number six. The output energizes the segments of the display that correspond to the number six.

Figure 5-4 shows the pinout for a 7447 seven-segment decoder. The data *input* lines correspond to the "weights" of the binary number. For example, the number 7 would be weighted as follows:

$$8 \quad 4 \quad 2 \quad 1$$
$$0 \quad 1 \quad 1 \quad 1$$
$$\text{MSB} \qquad \text{LSB}$$

The weights of the binary number can be written a different way:

$$2^3 \ 2^2 \ 2^1 \ 2^0$$

For any number, the least significant bit (LSB) is the number with the lowest weight, and the most significant bit is the number with the highest weight. Weights in binary numbers correspond to the powers of ten in our usual decimal numbering system. The rightmost digit is the number of 1s, the next to the left is the number of 10s, the next is the number of 100s. Notice how these follow the order: 10^0, 10^1, 10^2, etc.

The *outputs* of the decoder are wired to the segments with the corresponding letters.

The *lamp test* is normally in a high condition. If a logic 0 (0V) is delivered to this terminal, all of the segments of the display are lighted to test the lamp.

In some cases, it is desirable to eliminate zero from a count. This can be done by connecting the *blanking input* to common (logic 0, or zero volts).

A logic 0 delivered to the *blanking output* terminal turns off the display. That is useful for blanking the display during a count if it is only desired to show the final number.

Connect the display as shown in Fig. 5-3.

Note: In place of using switches, it may be more convenient to wire 5V to one strip of a terminal board and 0V to another strip. This is shown in Fig. 5-5. You can easily switch an input between logic 1 and logic 0 as shown by the examples in the illustration.

After changing the "switches" to show all of the displays, replace them with a counter as shown in Fig. 5-6. Use a one-second clock signal for the counter. Note that the display shows a count from 0 through 9, then starts with zero again.

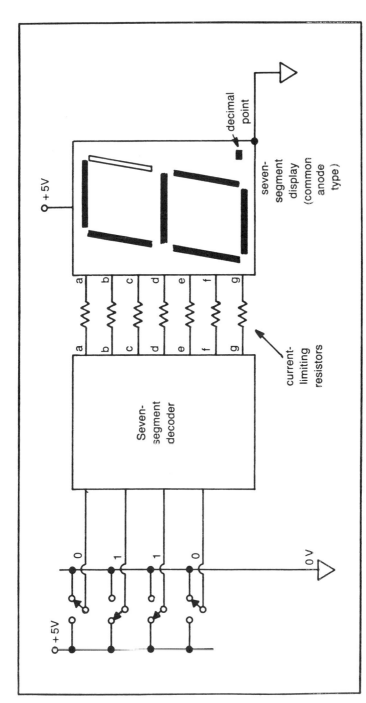

Fig. 5-3. The seven-segment decoder converts the input binary number (0110) to a lighted seven-segment display.

97

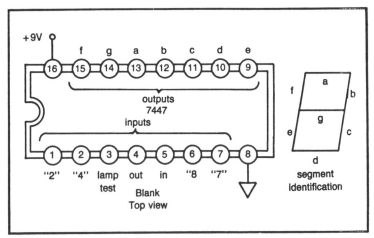

Fig. 5-4. The 7447 decoder is illustrated here. The lower case letters represent outputs to the seven-segment display. Each segment is identified in this illustration.

How can you use this setup to count *down* instead of *up*?

SELF TEST

(Answers at the end of the chapter.)

1. All memories can be divided into two basic categories: static and _____.

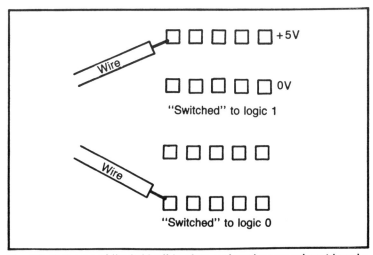

Fig. 5-5. This type of "switching" is often preferred on experiment boards.

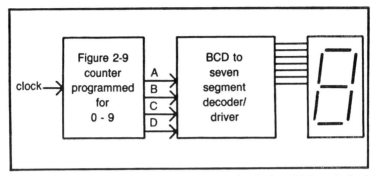

Fig. 5-6. With this counter and decoder system, the display will show decimal numbers. Program the counter for a 0-9 count.

2. You would expect a dynamic memory to be (volatile or non-volatile) because it utilizes a charged capacitor that must be refreshed.

3. Is this statement correct?: *A ROM cannot be randomly accessed.*

4. A flip-flop is used as a memory cell in a (static or dynamic) memory.

5. EPROMs are erased with _____ light.

6. For a CCD the charge representing memorized information is stored in a _____.

7. The bubbles in a bubble memory are actually shaped like a (sphere or cylinder).

8. Fusible links are used in _____.

9. A name sometimes used instead of RAM is _____/_____ memory.

10. Is this statement correct: An EPROM does not have to be removed from the circuit to be reprogrammed.

Answers to the Self Test

1. dynamic
2. volatile

3. The statement is NOT correct.
4. static
5. ultraviolet
6. potential well
7. cylinder (They look like circles, or bubbles, from the top.)
8. ROMs
9. read/write
10. The statement is NOT correct.

Chapter 6

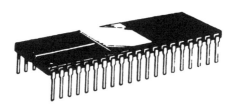

The RAM

In this chapter—

- Registers as memories
- Construction of a RAM
- Storing information into a RAM
- Getting information out of a RAM
- A RAM experiment

A microprocessor is an integrated circuit device that implements memory. So, studying memories helps to give a better understanding of how the microprocessor works.

Registers are memories inside the microprocessor. However, not all of them are called registers. Some examples of other names are: *accumulator* and *stack pointer*. Usually, registers can only store one or two bytes of memory.

RAMs store energy for a longer period of time. Also, they are capable of storing many bytes of data.

In Chapter 3 you saw how the registers were used during the run of a program. You also saw how a RAM was used to store the program. The concepts introduced in Chapter 2 will be extended in this chapter.

BINARY COMPONENTS

Any component that is stable in two states can be used as a

memory for binary numbers. Table 6-1 shows a few examples.

It is due to the fact that there are so many two-state devices that binary numbers are used in microprocessors and computers.

Either flip-flops or capacitors are used in registers and memories.

REVIEW OF SHORT-TERM MEMORIES

The registers in the theoretical microprocessor of Chapter 2 are actually used in the microprocessors of the real world. Each manufacturer uses its own terminology. For example, a *temporary register* of one manufacturer is called an *accumulator* by another.

The list of registers will now be reviewed and extended here. Always keep in mind the fact that registers are a form of memory used for short-term storage. Registers are also used in peripheral equipment.

Accumulators are registers that hold data for the ALU. They also temporarily hold data that is going into or coming out of the RAM.

The *instruction register* holds data for the instruction decoder. Op codes in a program are first loaded into the instruction register, then dumped into the decoder where the op code is interpreted and carried out.

A section of the RAM is set aside to store a sequence of data. This section is called the *stack*. It has a *last-in, first-out* (LIFO) organization.

The *stack pointer* is a register that holds the address for the top of the stack. That is where the first byte will come out of the stack.

The *condition code register* keeps track of various ALU operations. If you add 9 and 6 your get 15. If there are more digits, as

Table 6-1. Two-State Devices.

The Component	Logic 0	Logic 1
Switch	Open	Closed
Relay	Deenergized	Energized
Lamp	OFF	ON
Diode	Reverse Biased	Forward Biased
Flip Flop	Low	High
Magnetic Tape	Demagnetized	Magnetized
Capacitor	Discharged	Charged
Punched Tape	No Hole	Hole

in this simple problem:

$$\begin{array}{r} {}^{1}29 \\ \underline{36} \\ 5 \end{array}$$

you enter the 5 and carry the 1. When binary numbers are added it is also sometimes necessary to carry a 1. That carry is marked in the condition code register (CCR).

Another example of information stored in the CCR is whether an answer is positive or negative.

As your work with microprocessors continues you will learn about different methods of *addressing*. In other words, there are different methods used by the microprocessor to find memory addresses.

One method of addressing is called *indexing*. It finds the location in memory by combining the contents of an *index register* with another number. It may sound like a clumsy way to find an address, but it makes the microprocessor much more versatile.

The *program counter* is sometimes called a register because it holds the address of the next step in the program.

Some examples of registers used in microprocessors other than the 6800 are:

scratch pad. The scratch pad is a combination of registers used for temporary storage and manipulation of data.

status register. The status register is similar to the condition code register.

memory refresh register. The memory refresh register is used for refreshing dynamic memories which are discussed later in this chapter.

Microprocessors also utilize one-bit memories—called *flags*—to keep track of various microprocessor operations.

THE RAM COMPONENTS

Microprocessors and computers need large electronic storage places. They accomplish the same thing as registers, but they can store much more information. In fact, many types of RAMs—the name used for large storage places—are often made by using a large number of registers.

The term RAM (Random Access Memory) is misleading. It implies that you can access numbers in any address in the RAM (which is true) but you can't do that in other memories such as ROM (which is **not** true).

A more accurate term to describe this type of memory would be *read/write* (R/W). At one time there was a move to get people to use that expression, but that move seems to have died out.

Read/Write was thought to be a more accurate description because you can put information into the RAM and you can take information out of the RAM. You couldn't do that with the original ROM (Read Only Memory). Today, however, there are ROMs that *can* be programmed.

Probably the best way to compare a RAM with a ROM is to say that the RAM is *volatile*. In other words, when you shut off the power supply you lose the information that has been stored. This doesn't happen with ROMs, so they are called *non-volatile*.

There are special stand-by power supplies that will switch into the system very rapidly when the main supply fails. That saves the information in the RAM, but it doesn't make a RAM non-volatile. (If the RAM was really non-volatile, the stand-by power supply wouldn't be needed.)

Figure 6-1 shows two different ways to construct a RAM. In Fig. 6-1(a), a capacitor is charged to represent logic 1 and it is discharged to represent logic 0. The advantage of this type is that a large number of capacitors can be constructed on a single chip, so this type is known for the high amount of storage per integrated circuit.

Another advantage is that the information can be put in or taken out very rapidly which is another way of saying that it has a short access time.

The disadvantage of dynamic memories is that the capacitor can only hold its charge for a short period of time. To prevent loss of memory the capacitors must be recharged, or *refreshed*. This is usually accomplished with a special integrated circuit designed for the purpose.

Memories that are made with capacitors are called *dynamic RAMs*.

RAMs that are made with flip-flops—like the one in Fig. 6-1(B) are called *static RAMs*. They do not require a refresh voltage to keep them in the proper storage condition.

The experiments that you will perform for this chapter will use static RAMs.

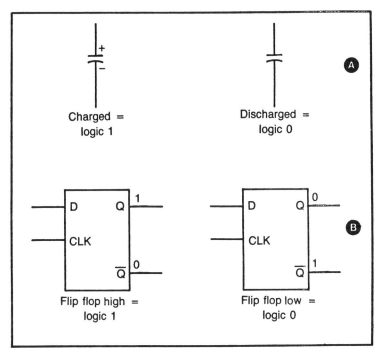

Fig. 6-1. Charged or discharged capacitors (A) are used for dynamic memories. D flip flops (B) are used for static memories.

A Typical RAM

If you were a microprocessor there would be certain control signals that you would have to deliver to a RAM before you could put information *in* (write) or take information *out* (read). To keep those terms straight in your mind, remember that you "write in," but you "read out."

The operating signals are delivered on a control bus. Figure 6-2 summarizes them. You may not find all of these control inputs on some particular RAM, but you should know them, because when you are working with different RAMs over a period of time, you will see each signal.

When there is an overbar with a control input it means that it requires an active low.

For example, CS means that the Chip will be Selected when a logic 1 is delivered to that terminal. In other words, it requires an *active high.*

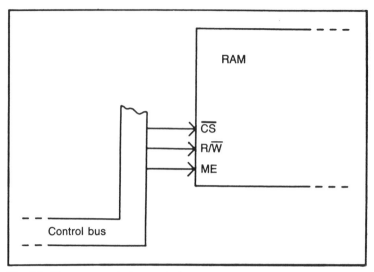

Fig. 6-2. These are typical controls for a memory.

However, $\overline{CS}$ is also a Chip Select terminal. It means that the chip will be selected when a logic 0 is delivered to that terminal. So, it requires an *active low*.

Additional control inputs and their meanings will now be discussed.

R/$\overline{W}$. This is a *read/write* input. Note that an active low is required to write information into memory. A logic 1 delivered to this terminal is required if you want to read out information that has been stored. If you want to store information in the memory the terminal must receive a logic 0.

The required signal must be delivered continuously while the information is going in or coming out.

ME—Memory Enable. This terminal requires a logic 1 while the memory is being used, and a logic 0 when it is in a standby condition.

If the memory didn't have this terminal, or one that performs the same function, it could deliver undesired signals to the data bus. Chip Enable (CE) does the same thing as memory enable.

CS—Chip Select. This input does the same thing as the memory enable but it is sometimes used in a different way. There may be several RAMs available to the microprocessor. It picks the desired one by delivering a logic 1 to this terminal.

A decoder may be utilized so that a binary number will select the desired memory address. The decoder would permit the use

106

of 16 different memories by using a code having only four binary digits.

Figure 6-3 shows the block diagram of a 64-bit static RAM. It stores 16 words, each having a word length of four bits. Four bits is usually called a *nibble*.

Four lines deliver the address selection. This is a binary number input, so 0010 will select one line of the memory.

Figure 6-4 gives a better idea of how the addresses are decoded. This example has only four lines. Each word is four bits wide. Note that (as shown by the circled numbers) a binary 10 input on the address bus selects line 2, which is the second from the bottom.

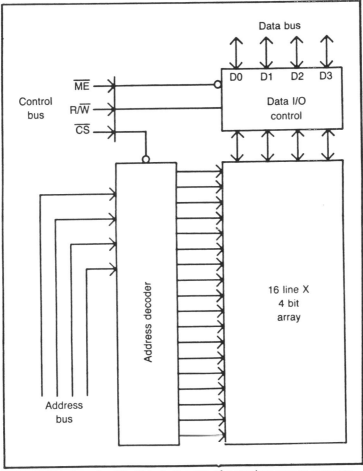

Fig. 6-3. This is a simplified block diagram of a random access memory.

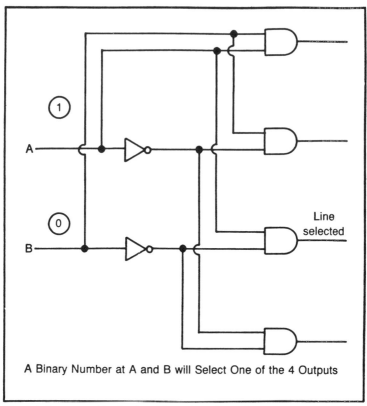

Fig. 6-4. Construct this simple address decoder to demonstrate how memory addresses are selected by the program counter.

Returning to Fig. 6-2, the data bus delivers data in and carries data away—depending upon the setting of the R/$\overline{\text{W}}$ control input.

In some cases you will see a DI for *data in* and $\overline{\text{DO}}$ for *data out*. This means that the output will be the compliment of the data that was stored. So, if 1101 was stored, the output will be 0010 when the data is retrieved. With this type of memory there is a way to avoid confusion. Use inverters at the input. This automatically stores the compliment of the input, so that the output will be the desired value.

PROGRAMMED REVIEW

(See the instructions for the Programmed Review in Chapter 1 for using this section.)

Block 1

Which of the following is usually considered to be volatile?

(a) ROM *Go to Block 8.*
(b) RAM *Go to Block 12.*

Block 2

Your answer to the question in Block 21 is not correct. Read the question again, then *go to Block 4.*

Block 3

The correct answer to the question in Block 18 is (a). Some memories have an ME terminal that is used to turn the I.C. on.
Here is your next question: The output of the address decoder is

(a) one digit for any input binary number. *Go to Block 19.*
(b) a binary coded decimal number. *Go to Block 14.*

Block 4

The correct answer to the question in Block 21 is (b). The registers in I/O will be discussed later. However, it was clearly stated in this chapter that those registers do exist.
Here is your next question: A memory that does *not* lose its stored information when the power supply voltage is lost is said to be

(a) non volatile. *Go to Block 22.*
(b) static. *Go to Block 9.*

Block 5

Your answer to the question in Block 12 is not correct. Read the question again, then *go to Block 10.*

Block 6

Your answer to the question in Block 19 is not correct. Read the question again, then *go to Block 24.*

Block 7

Your answer to the question in Block 19 is not correct. Read

the question again, then *go to Block 24.*

Block 8

Your answer to the question in Block 1 is not correct. Read the question again, then *go to Block 12.*

Block 9

Your answer to the question in Block 4 is not correct. Read the question again, then *go to Block 22.*

Block 10

The correct answer to the question in Block 12 is (a). Dynamic memories require periodic refresh voltage to recharge the capacitors.

Here is your next question: A designation of CS means

(a) Chip Select. It is a *control* terminal. *Go to Block 18.*
(b) Chip Storage. It is used as a terminal for storing *data* in the memory. *Go to Block 25.*

Block 11

Your answer to the question in Block 18 is not correct. Read the question again, then *go to Block 3.*

Block 12

The correct answer to the question in Block 1 is (b). Information stored in a RAM will be lost when the power goes off unless a standby supply is used.

Here is your next question: The charge or discharge of a capacitor represents a 1 or 0 in a

(a) dynamic memory. *Go to Block 10.*
(b) static memory. *Go to Block 5.*

Block 13

Your answer to the question in Block 24 is not correct. Read the question again, then *go to Block 21.*

Block 14

Your answer to the question in Block 3 is not correct. Read the question again, then *go to Block 19.*

Block 15

Your answer to the question in Block 24 is not correct. Read the question again, then *go to Block 21.*

Block 16

Your answer to the question in Block 22 is not correct. Read the question again, then *go to Block 20.*

Block 17

Your answer to the question in Block 24 is not correct. Read the question again, then *go to Block 21.*

Block 18

The correct answer to the question in Block 10 is (a). This terminal is for control—not data.

Here is your next question: Which of the following does the job of a memory enable (ME)?

(a) CE *Go to Block 3.*
(b) DE *Go to Block 11.*

Block 19

The correct answer to the question in Block 3 is (a). If the memory has 16 addresses then the address will be selected by a 1 of 16 decoder.

Here is your next question: Which of the following might be the data output of a RAM when data stored was 1001?

(a) 1111 *Go to Block 6.*
(b) 0110 *Go to Block 24.*
(c) Depending upon the way the RAM is made, either of the two outputs could be obtained. *Go to Block 7.*

Block 20

The correct answer to the question in Block 22 is (b). This designation is for a control signal that determines if information is coming out of or going into memory.

Here is your next question: A certain control terminal requires a logic 0 for operation. In other words, the terminal requires an _____ _____. *Go to Block 23.*

Block 21

The correct answer to the question in Block 24 is (c). The read (out) terminal must be at logic 1. The $\overline{E}$ terminal requires an active low.

Here is your next question: Which of the following statements is correct?

(a) You would NOT expect to find registers in the I/O section of a microprocessor system. *Go to Block 2.*
(b) You would expect to find registers in the I/O section of a microprocessor system. *Go to Block 4.*

Block 22

The correct answer to the question in Block 4 is (a). Don't confuse the term *static*, which means the memory doesn't have to be refreshed, with *non-volatile*, which means the stored information doesn't depend upon the power supply voltage.

Here is your next question: The designation R/$\overline{W}$ means

(a) the memory can read but not write. *Go to Block 16.*
(b) the input must be logic 1 for read and logic 0 for write. *Go to Block 20.*

Block 23

The correct answer to the question in Block 20 is *active low*.

Here is your next question: Another name for accumulator is _____. *Go to Block 26.*

Block 24

The correct answer to the question in block 19 is (b). If the stored data was 1001 then the output data should be either 1001 or 0110—depending upon whether the data is inverted. There is no way that the output data could be 1111.

Here is your next question: Which of the combinations in the

112

following table will set up the memory to deliver stored information to the outside world?

	R/$\overline{\text{W}}$	$\overline{\text{E}}$
Condition 1:	0	0
Condition 2:	0	1
Condition 3:	1	0
Condition 4:	1	1

(a) The one in the row marked condition 1. *Go to Block 13.*
(b) The one in the row marked condition 2. *Go to Block 17.*
(c) The one in the row marked condition 3. *Go to Block 21.*
(d) The one in the row marked condition 4. *Go to Block 15.*

Block 25

Your answer to the question in Block 10 is not correct. Read the question again, then *go to Block 18.*

Block 26

The correct answer to the question in Block 23 is *register.* You have now completed the programmed review.

EXPERIMENTS

In the experiments for this chapter you construct some of the basic circuits used in a microprocessor system.

The Address Decoder

Construct the 1-of-4 decoder shown in Fig. 6-4. With the binary counter sequencing from 00 through 11 the LEDs should light in sequence.

In this experiment you have simulated the address decoder. The binary counter does the job of the program counter, but the program counter actually sequences through the address where the program is stored and then stops at the last step.

The Memory

Connect the binary (0000 to 1111) counter of Fig. 2-9 to the address input of the 7489 RAM as shown in Fig. 6-5. (Details of the 7489 RAM are given in the Appendix.) Use the bounceless switch in place of the clock to sequence the counter one address at a time. After you select a memory location set the data switches.

The LEDs on the counter output tell you what address line you are on. Remember that you count 0000 as the first address.

Load the memory so that the first four addresses contain codes that will light the LEDs connected to the data output in a one-at-a-time sequence pattern. In other words, for the first four addresses in memory you will store the following codes:

Address	Stored Code
0000	0001
0001	0010
0010	0100
0011	1000

If you repeat the stored code for the next four addresses, the LEDs will sequence again. If you repeat the code four times, the LEDs will sequence continuously. After the highest count is reached (address 1111) the counter will reset to 0000, so the sequence should continue.

Replace the bounceless switch with the counter clocked at approximately 1 hertz. Change the instruction from write to read, and observe that the LEDs sequence.

Change the stored numbers in the memory so that the LEDs strobe back and forth. Your first eight addresses should have the following data:

Address	Stored Code
0000	0001
0001	0010
0010	0100
0011	1000
0100	0100
0101	0010
0110	0001
0111	0000

Connect the seven segment decoder and display in place of the output LEDs as shown in Fig. 6-6.

114

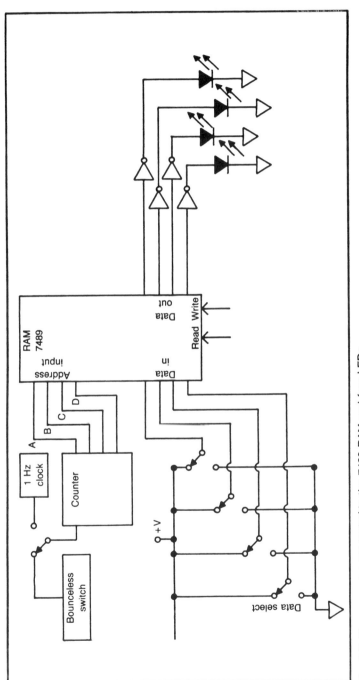

Fig. 6-5. Binary counter connected to the 7489 RAM and four LEDs.

115

116

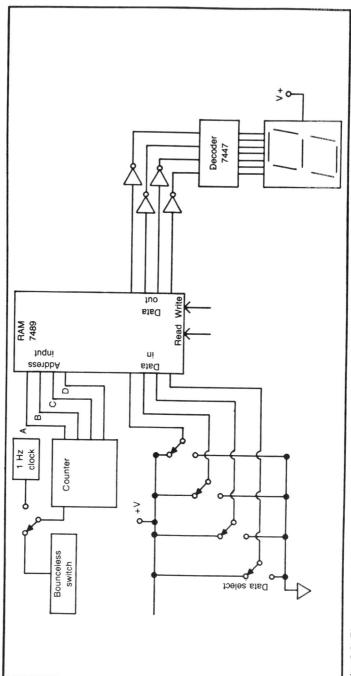

Fig. 6-6. Binary counter connected to the 7489 RAM and a seven segment display.

Store your telephone number into the first seven addresses of the RAM. Use the single-step bounceless switch when you are storing the binary equivalent of your phone number.

After you have stored your phone number, use only the three least significant bits of the counter to sequence the addresses from 000 to 111.

The seven-segment display should count through your phone number in sequence.

You can insert a time space between the last number and first number by using all four flip-flops of the counter. The decoder will turn off all LEDs in the display when it receives an input of 1111, so store that binary number in *memory* addresses 1000 and 1001.

Wire the counter so that it counts from 0000 to 1001 and then recycles.

Now your phone number will display, followed by a pause, and then display again.

Add your area code to the display of your phone number.

SELF TEST

(Answers at the end of the chapter)

1. Which of the following is NOT a control signal that the microprocessor might send to a microprocessor?

   (a) CE
   (b) D0
   (c) CS
   (d) R/$\overline{W}$

2. Dynamic memories are made with

   (a) inductors.
   (b) resistors.
   (c) transformers.
   (d) capacitors.

3. The address bus delivers a binary count to the address

   (a) decoder.
   (b) encoder.
   (c) counter.
   (d) inverter.

4. The stack is located in

(a) a ROM.
(b) a RAM.
(c) the ALU.
(d) in the address decoder.

5. The index register is used for

(a) logic comparison.
(b) display.
(c) addressing.
(d) decoding.

6. Which of the following is a function of the condition code register?

(a) Hold information about whether an ALU answer is positive or negative.
(b) Hold the location of the stack.
(c) Hold the location of the next step in a program.
(d) Decode an address.

7. In some microprocessor systems a combination of registers used for temporary storage is called a

(a) tablet.
(b) folder.
(c) scratch pad.
(d) bin.

8. Which of the following is the name given for a one-bit memory used to keep track of microprocessor operations?

(a) Sign
(b) Flag
(c) Fence
(d) Basket

9. Which of the following is a descriptive name sometimes given to a RAM?

(a) I/O

(b) ON/OFF

(c) Q/$\overline{Q}$

(d) read/write

10. One thing that really distinguishes a RAM from a ROM is that the RAM

(a) can be randomly accessed.

(b) is non volatile.

(c) is volatile.

(d) has a data output capability.

Answers to the Self Test

1. (b)
2. (d)
3. (a)
4. (b)
5. (c)
6. (a)
7. (c)
8. (b)
9. (d)
10. (c)

Chapter 7

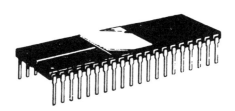

Interfacing

In this chapter—

- Why interfacing circuits and components are needed
- What an Asynchronous Communications Interface Adapter does
- How the two-way bus works
- What a Peripheral Interface Adapter does
- Experiments with interfacing

THE NEED FOR INTERFACING

Interfacing is used in electronic systems when the output of one stage cannot be directly coupled to another stage. Figure 7-1 shows some examples of how interfacing is used.

For example, if the output of a CMOS logic gate is delivered to an ECL gate it is necessary to use an intermediate coupling device (interface) to match the positive CMOS voltage to the negative ECL voltage.

Interfacing is also needed to match an eight-line bus to a single telephone line.

Most microprocessors deliver a five-volt code to the outside world. That five volts is insufficient to run many of the peripheral devices that the microprocessor is used to control.

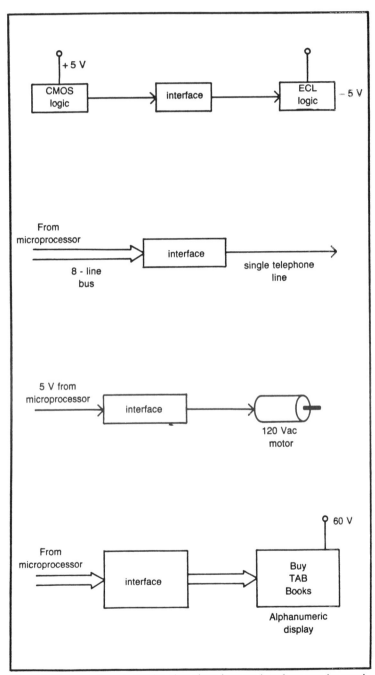

Fig. 7-1. Here are four examples that show how an interface can be used.

As an example, if you want to operate a motor, and also control the speed of that motor, the five volts delivered by the microprocessor will not be sufficient for the job.

Suppose, on the other hand, you want to use the microprocessor to operate a 60-volt alphanumeric display. Again, the microprocessor would not be able to deliver sufficient voltage.

What you would need to run and control the motor, or to operate an alphanumeric display, is an interface between the microprocessor and the outside world. The interface converts the low-power five-volt output of the microprocessor system to the higher voltage and power for operating peripheral devices.

When data is delivered *to* the microprocessor it must also be at the five-volt level. You cannot deliver any other voltage value to operate the microprocessor data input.

So, an interface makes it possible to link the microprocessor with the outside world.

Interfaces are not just used in logic and microprocessor systems. They are also used in linear circuits.

THE TRANSDUCER AS A DATA SOURCE

Suppose the microprocessor is to receive data from an external *transducer*. By definition, a transducer is a device that permits the energy from one system to control the energy of another system. Sometimes, for convenience, we say that the transducer "converts" energy from one form to another.

To be scientifically correct, transducers cannot convert energy from one form to another. They simply control one type of energy with another type of energy. A loudspeaker is one example of a transducer. Electrical energy in the speaker *controls* the sound energy. Another example is a microphone in which sound energy *controls* electrical energy.

Many different types of transducers are used in industrial electronic control systems. They are also called *sensors*.

The microprocessor must be operated from a wide variety of voltages and currents. If you have any input from a transducer that is not five volts you will need an interface. Also, the microprocessor must deliver electrical signals to operate devices that require a wide variety of voltages and currents for their operation.

THE OPTICAL COUPLER

There are a number of integrated circuits available for matching logic families. Also, buffers are available for matching unequal voltages and unequal powers. You can find these specialized ICs in catalogs. There is such a wide variety of these specialized that it wouldn't be possible to list them in a book this size. So, we will not discuss them in this chapter.

One interfacing device that can be used in a wide variety of interfacing jobs is shown in Fig. 7-2. It is called an *optical coupler*. As shown in the illustration, there is a wide range of types available.

Consider the simple type shown in Fig. 7-3. The input side has a light-emitting diode (LED); and, the output side has a light-activated diode (LAD).

The simple circuit of Fig. 7-3(A) utilizes a low-power five-volt circuit to operate a 24V lamp.

When there is no light being emitted by the LED the LAD has a very high impedance. In fact, it is like an open circuit.

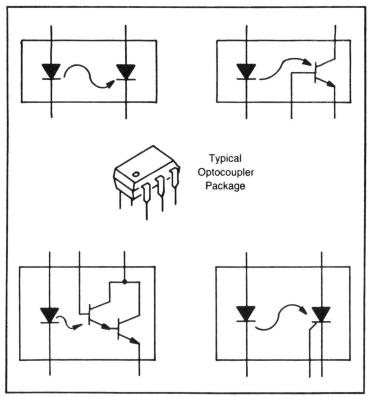

Fig. 7-2. Four types of optical couplers (also called optocouplers) are shown here.

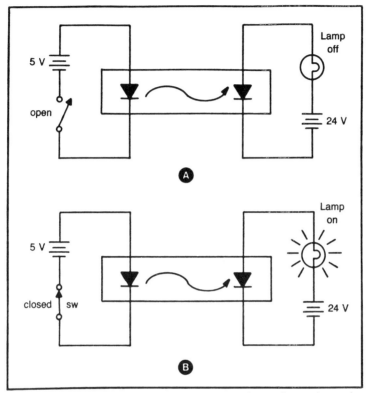

Fig. 7-3. This simple circuit shows how an optocoupler can be used as an interface between a low voltage control and a high voltage lamp. (A) The switch is open so the LED is not lit. Therefore, there is no coupling between the input and output circuits. (B) With the switch closed, the LED is on. This produces coupling between the two circuits and the lamp is on.

If a voltage is applied to the LED, it produces light. That light activates the LAD and lowers its impedance to the point where it can be disregarded in the forward direction.

As shown in Fig. 7-3(B), the five-volt input circuit is sufficient to turn on the LED, and the 24V lamp is turned on by the forward-conducting LAD.

One reason for the popularity of the optical coupler is the very high impedance between the input and output circuits. For practical applications the two circuits can be considered to be completely isolated from each other.

Versatility is another reason for their popularity. The same type of optical coupler can be used to match many different voltage and power levels. Also, they are available in a convenient IC package.

PARALLEL-TO-SERIAL INTERFACING

The data output of a microprocessor is usually delivered on a four-line, eight-line or sixteen-line bus. All of the bits that make up the code for each word are delivered simultaneously—one line of the bus being used for each bit. This is called a *parallel output*.

In some applications the microprocessor is used to send signals to some distant location over a telephone line. If you tried to do that with the parallel output directly you would need 8 telephone lines—one for each line on the data bus.

Eight long distance telephone lines would be very expensive. To get around this problem, we use *parallel-to-serial converters*. In the case of the 6800 family, there is an integrated circuit specifically designed for this purpose. It is called the *Asynchronous Communications Interface Adapter* (ACIA). Figure 7-4 illustrates its operation.

The ACIA takes the eight bits of parallel data for each word from the microprocessor and converts it into a series of pulses that can be sent on a single telephone line. Sending a word one bit at a time is called *serial transmission*.

A parallel-in/serial-out register is used in the ACIA to accomplish the data conversion.

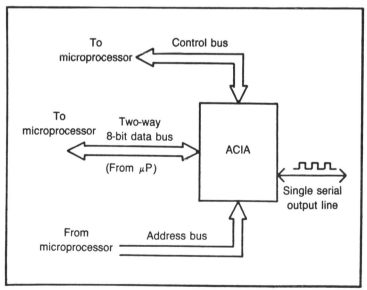

Fig. 7-4. The ACIA converts information on the 8-bit data bus into a serial output. Also, a serial input can be converted to parallel information on the 8-bit data bus.

126

It is reasonable to expect that a microprocessor in some remote location will be receiving the information being sent on the telephone line in serial form. The microprocessor itself cannot work directly with that serial data. All data must be delivered to the microprocessor on a parallel bus.

So, some method must be used to convert the serial pulses into parallel. This is also accomplished with the ACIA. A *serial-to-parallel* register is utilized.

A block diagram of the ACIA and a more complete description of its operation are given in the Appendix.

Many computers deliver their output information in a parallel format similar to that of the microprocessor. As with the microprocessor, the computer can be used to link up with distant stations on a telephone line. To do this a special interface is needed. It is called a *modem* (**mod**ulate/**dem**odulate). The modem permits a computer to correspond directly with another computer at some distance. It is necessary that both computers be operating with the same type of modem.

THE PIA

Many of the peripherals that receive instructions from the microprocessor can work directly with its parallel output. However, an interface is still recommended by the manufacturer. In the case of the 6800 the interface is called a Peripheral Interface Adapter (PIA). Manufacturers of other microprocessors call their version Peripheral Input/Output (PIO).

The PIA acts like a traffic controller for input and output signals. It is designed to interface with two different peripheral bus lines.

Figure 7-5 shows a simplified drawing of the PIA functions. A detailed block diagram is given in the appendix along with an explanation of its operation. Observe the use of parallel in/parallel out busses used for its operation.

The data terminals on the microprocessor are used for both input and output. The traditional way to handle two-way traffic on a bus line is to use a two-way bus. An example is shown in Fig. 7-6.

The AND gates are used as ENABLEs, so that when there is a logic 1 on one terminal, whatever is on the other terminal is passed through the gate. By using an inverter and a buffer at the ANDs enable input terminals it is impossible to enable both ANDs at the same time.

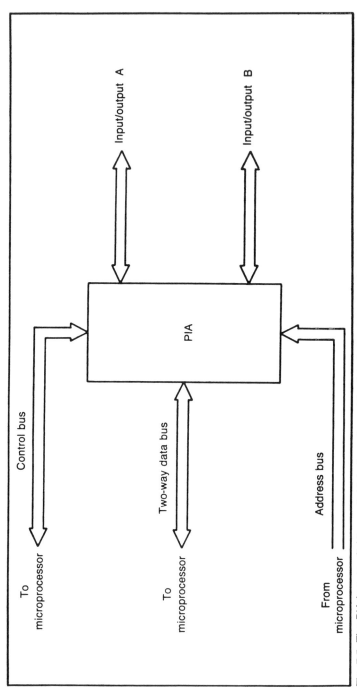

Fig. 7-5. The PIA is a parallel-to-parallel interface. It provides interfacing between the microprocessor and two peripheral systems.

128

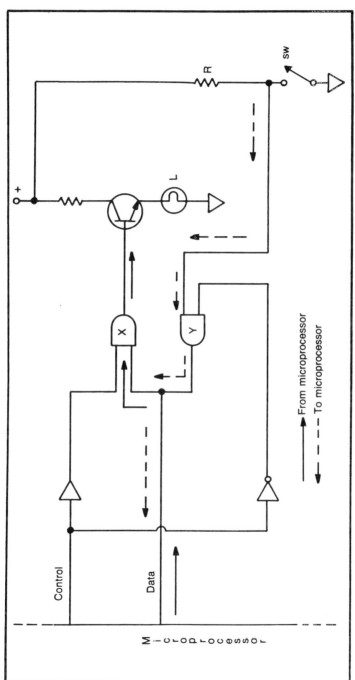

Fig. 7-6. This is one example of a circuit for a two-way bus.

→ From microprocessor
--→ To microprocessor

The line marked *data* is the actual two-way bus. When there is a logic 1 on the control line the AND gate marked X is enabled. The microprocessor delivers a logic 1 (+5V) through X to the base of the transistor. That, in turn, biases the transistor into conduction and the lamp is ON.

When there is a logic 0 (0V) on the control line Gate Y is enabled. In that case, either a logic 0 or logic 1 will be delivered to the microprocessor through Y—depending upon whether the switch is open or closed.

The PIA has a *data direction register* (DDR) that serves the same purpose as the control line in the circuit of Fig. 7-6.

MEMORY MAPPING

The thing that determines whether or not information is going between the microprocessor and the outside world is the program. There are two ways that the interchange can take place. In the 8080 microprocessor, and other types, there is a special code for delivering or receiving data through the PIO.

For the 6800 and other microprocessors the complete system is *memory mapped*. That means the peripheral interface devices are considered to be memory locations, so to deliver data from the microprocessor through the PIA, the programmer simply calls out a memory address which is actually the PIA.

Likewise, the programmer can send information through the ACIA by delivering it to a memory address. In reality, that memory address is the ACIA, which is actually that interface. Memory mapping simplifies the use of peripheral interfacing.

PROGRAMMED REVIEW

(See the instructions for the Programmed Review in Chapter 1.)

Block 1

When the interface device has a memory address the system is said to

(a) have a memory call-down. *Go to Block 8.*
(b) be memory mapped. *Go to Block 12.*

Block 2

The correct answer for the question in Block 18 is (b). The example of a two-way bus given in this chapter is not the only way it can be done. Each designer has a preference.

Here is your next question: How is the direction of data through the PIA determined?

(a) By the DIO. *Go to Block 24.*
(b) By the DDR. *Go to Block 15.*

Block 3

Your answer to the question in Block 7 is not correct. Read the question again, then *go to Block 14.*

Block 4

The correct answer for the question in Block 15 is (b). There are two kinds of transducers, or sensors. *Active* types generate a voltage that is proportional to what is being sensed. *Passive* types change resistance, capacitance or inductance in accordance to what is sensed.

Here is your next question: The parallel interface for the Motorola microprocessor is called PIA. For other manufacturers it is sometimes called

(a) PIO. *Go to Block 23.*
(b) PINO. *Go to Block 17.*

Block 5

The correct answer for the question in Block 23 is (b).

Here is your next question: A device that allows the energy of one system to control the energy in another system is called _____. *Go to Block 26.*

Block 6

The correct answer for the question in Block 19 is (b). You could also use one of the specialized integrated circuits designed for this job.

Here is your next question: Which of the following is used as a voltage interface?

(a) Optical coupler. *Go to Block 22.*
(b) PIA. *Go to Block 16.*

Block 7

The correct answer for the question in Block 22 is (a). A typical value of isolation resistance is 10^{12} ohms.

Here is your next question: To transmit data on a single telephone line you would need

(a) serial transmission. *Go to Block 14.*
(b) parallel transmission. *Go to Block 3.*

Block 8

Your answer to the question in Block 1 is not correct. Read the question again, then *go to Block 12.*

Block 9

Your answer to the question in Block 15 is not correct. Read the question again, then *go to Block 4.*

Block 10

Your answer to the question in Block 19 is not correct. Read the question again, then *go to Block 6.*

Block 11

Your answer to the question in Block 18 is not correct. Read the question again, then *go to Block 2.*

Block 12

The correct answer for the question in block 1 is (b). Some experts claim that all microprocessors are memory mapped. It is simply a matter of how you define memory mapped.

Here is your next question: Computers can exchange information on a telephone line by using a

(a) transducer. *Go to Block 21.*
(b) modem. *Go to Block 19.*

Block 13

Your answer to the question in Block 23 is not correct. Read the question again, then *go to Block 5.*

Block 14

The correct answer for the question in Block 7 is (a). Serial transmission is also used for putting data on cassette tapes.

Here is your next question: The integrated circuit used to convert a parallel input to a serial output is called the

(a) PIO. *Go to Block 25.*
(b) ACIA. *Go to Block 18.*

Block 15

The correct answer for the question in Block 2 is (b). You should take some time to study the PIA and ACIA block diagrams and explanations given in the Appendix.

Here is your next question: Another name used for transducer is

(a) input/output. *Go to Block 9.*
(b) sensor. *Go to Block 4.*

Block 16

Your answer to the question in Block 6 is not correct. Read the question again, then *go to Block 22.*

Block 17

Your answer to the question in Block 4 is not correct. Read the question again, then *go to Block 23.*

Block 18

The correct answer for the question in Block 14 is (b). ACIA is the term Motorola uses for the parallel-to-serial interface.

Here is your next question: For the two-way bus described in this lesson the AND gates are connected as

(a) BUFFERS. *Go to Block 11.*
(b) ENABLES. *Go to Block 2.*

Block 19

The correct answer for the question in Block 12 is (b). There are companies that specialize in making modems for computers.

Here is your next question: In order to deliver the output of a CMOS gate to an ECL input you could use

(a) a frequency converter. *Go to Block 10.*
(b) an optical coupler. *Go to Block 6.*

Block 20

Your answer to the question in Block 22 is not correct. Read the question again, then *go to Block 7.*

Block 21

Your answer to the question in Block 12 is not correct. Read the question again, then *go to Block 19.*

Block 22

The correct answer for the question in Block 6 is (a). The PIA works with the same input and output voltage.

Here is your next question: Which of the following is an important advantage of an optical coupler?

(a) High isolation impedance. *Go to Block 7.*
(b) Excellent frequency response. *Go to Block 20.*

Block 23

The correct answer for the question in Block 4 is (a).

Here is your next question: In order to use a microprocessor for operating a 60-volt alphanumeric display, you could employ

(a) a PIA. *Go to Block 13.*
(b) an optical coupler. *Go to Block 5.*

Block 24

Your answer to the question in Block 2 is not correct. Read the question again, then *go to Block 15.*

Block 25

Your answer to the question in Block 14 is not correct. Read the question again, then *go to Block 18.*

134

Block 26

The correct answer to the question in Block 5 is *transducer* (or *sensor*). You have now completed the programmed review.

EXPERIMENTS

In the experiments for this chapter you will construct and operate some interface circuitry.

The Two-Way Bus

Construct the two-way bus circuit of Fig. 7-6. Show that information can be passed on the data line in either direction, depending upon whether the control input is at logic 1 or logic 0. In place of the output load you can use an LED to demonstrate that the circuit is working.

It is possible to construct a two-way bus using the same principle as in Fig. 7-6 but using a quad NAND integrated circuit. The ENABLE circuits may require additional inverters.

Optical coupler

Figure 7-7 shows a typical application for a coupler that utilizes a photodarlington.

Construct this circuit. If you cannot obtain the specific parts shown, use an optical coupler to control any output load.

It is important to understand that an LED has a voltage across it of 1.5 volts (approximately) when it is in the ON condition. Since you can't see inside the optical coupler, one way to determine if the LED is on is to measure the voltage across its terminals.

SELF TEST

(Answers at the end of the chapter)

1. When the microprocessor considers the interfacing devices as being located in a memory address, the system is said to be
   (a) memory located.
   (b) memory interfaced.
   (c) memory mapped.
   (d) address in memory.

2. A transducer, or sensor, that generates a voltage related to

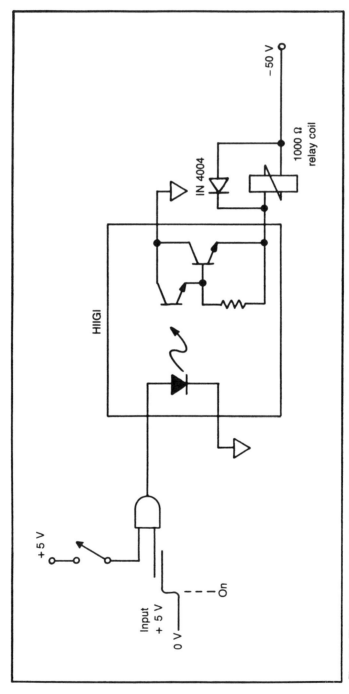

Fig. 7-7. This is a manufacturer's (Motorola, Inc.) suggested interface circuit. It permits a TTL logic output to control a 50-volt, 1,000 ohm relay.

the quantity sensed is called
(a) an active transducer.
(b) a passive transducer.
(c) a resistive transducer.
(d) None of these choices is correct.

3. Which of the following is an example of a device that can be used as a voltage interface?
(a) PIA
(b) ACIA
(c) PIO
(d) Optical Coupler

4. Computers can signal each other over a telephone line by using
(a) optical couplers.
(b) modems.
(c) transducers.
(d) relays.

5. Which of the following can be used to record digital data on a cassette tape?
(a) PIA
(b) PIO
(c) ACIA
(d) ACEF

6. Which of the following gates is used to make a two-way bus for the experiment in this chapter?
(a) Zener diodes
(b) Relays
(c) EXCLUSIVE OR gates
(d) AND gates

7. Which of the following can be used to couple the output of ECL logic output to a TTL system?
(a) PIO

(b) ACIA
(c) Optical Coupler
(d) PIA

8. An advantage of an optical coupler as a voltage interface is that it has
   (a) no voltage requirement.
   (b) no current requirement.
   (c) only two terminals.
   (d) high isolation impedance.

9. To operate a 60-volt alphanumeric display by using a microprocessor you could use
   (a) a PIA.
   (b) an optical coupler.
   (c) a PIO.
   (d) an ACIA.

10. A data direction register is used in the
    (a) PIA.
    (b) PIO.
    (c) ACIA.
    (d) None of these answers is correct.

Answers to the Self Test

1. (c)
2. (a)
3. (d)
4. (b)
5. (c)
6. (d)
7. (c)
8. (d)
9. (b)
10. (a)

Chapter 8

Putting the
Microprocessor to Work

In this chapter—

- passive transducers
- active transducers
- D/A and A/D converters
- phase-locked loops
- microprocessor applications

WHAT ARE THE SOURCES
OF MICROPROCESSOR INFORMATION?

In order for the microprocessor to do its job it must have input information in the form of digital data. Some of the sources for that data have already been discussed but they will be presented in greater detail here.

One obvious way to get information into a microprocessor is shown in Fig. 8-1. Mechanical switches are good binary sources. They can present a logic 0 when they are open and a logic 1 when they are closed.

Mechanical switches are not normally used for direct inputs to microprocessors without accompanying circuitry. One problem is that the contacts in the switch have a tendency to bounce and that causes false inputs. To get around this problem, simple de-

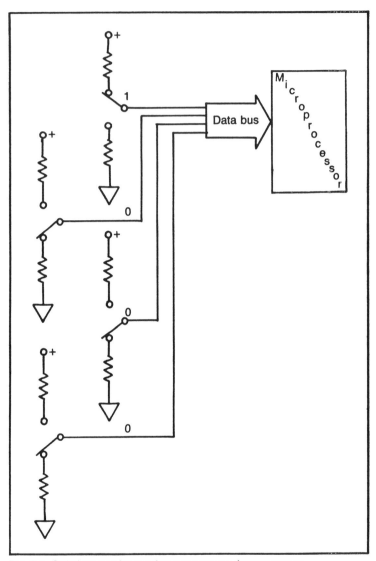

Fig. 8-1. Switches can be used to program a microprocessor.

bouncer circuits like the one shown in Fig. 8-2 are used to clean up the switch output.

It would be possible to completely program a microprocessor by using switches. As a matter of fact this was done in some early computer systems. The switches were used to deliver a binary code, with one switch used for each of the input lines on the data bus.

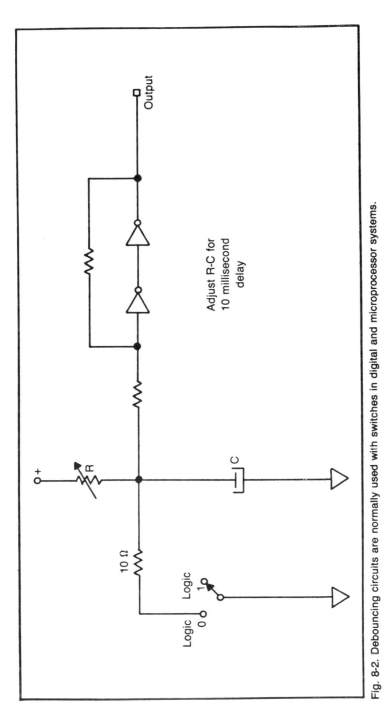

Fig. 8-2. Debouncing circuits are normally used with switches in digital and microprocessor systems.

After the information for one word was dumped into the microprocessor the bus line was opened and the switches were again set for the next input.

You wouldn't be able to do this many times before you became very tired of the procedure. If you were trying to put a long program into the microprocessor the use of switches would be very time consuming and very tiresome.

A step up from using switches is the keyboard. Actually each key on the keyboard is a switch representing an alphanumeric character identified by the letter on the keyboard. A good example of an alphanumeric keyboard is a typewriter.

The letters and numbers on the keyboard are connected to switches. The switch output goes to an *encoder* as shown in Fig. 8-3. The encoder converts that switch contact into binary codes to be delivered to the microprocessor.

Decoders, as you might expect, convert binary codes back into the equivalent of an On-Off switch for operating some peripheral device.

In addition to switches and keyboards, there are other On-Off devices that are convenient for putting information into the microprocessor. Two of these are shown in Fig. 8-4.

A shaft encoder is illustrated in Fig. 8-4(A). It has combinations of dark and light areas on a disc connected to a motor shaft. Ideally, a stepping motor is used for this operation. As the motor rotates, the light shines through (or, reflects from) the light and dark areas onto light sensitive pick-ups. The output of these pick-ups is a binary code corresponding to shaft position.

A *bar code,* shown in Fig. 8-4(B), provides a binary input to the microprocessor through a reading wand. The wand is moved across the code and the resulting binary input from the light and dark areas is processed and delivered on the microprocessor bus line.

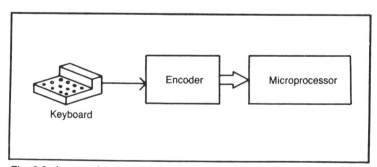

Fig. 8-3. An encoder converts the key switch to a binary code.

Fig. 8-4. Examples of microprocessor coding devices. (A) A shaft encoder. (B) An example of a bar code.

Passive Transducers

In many applications the microprocessor receives its information from external transducers. This type of input was mentioned briefly in Chapter 7 and will be extended here.

There are two kinds of transducers: *passive* and *active*. In electronics active devices deliver a voltage (or current) that is proportional to a parameter being sensed. Passive transducers affect the resistance, capacitance, and/or inductance of a circuit—but they do not produce voltage. We will look at a passive transducer first.

Figure 8-5 shows an application of a passive transducer in the form of resistance. In each case a resistor produces a change in circuit resistance that corresponds to what is being sensed. Keep in mind the fact that this (and those that follow) is a simplified diagram. In operation the devices are usually more sophisticated than shown here.

A *thermistor* is a non-linear resistor. Its resistance value changes over a wide range of values for relatively small changes in temperature. Thermistors are normally used for sensing temperatures or temperature changes.

In most applications the thermistor sensor is placed in one leg of a bridge circuit. The other leg often contains an identical thermistor that is used as a reference. At room temperature the inputs from both thermistors should be the same and the bridge should

143

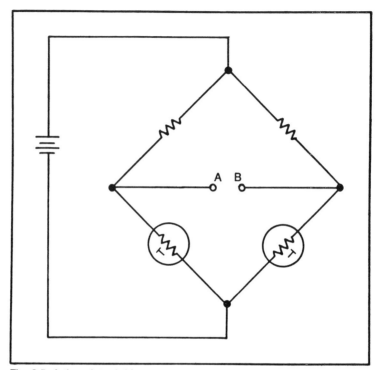

Fig. 8-5. A thermistor bridge.

be balanced. In other words, there should be no voltage across points A and B in the circuit of Fig. 8-5.

When the thermistor is placed in the location where temperature is to be sensed, its resistance is changed because of that temperature. The resulting resistance change causes the bridge circuit to be unbalanced in an amount that is directly related to the sensed temperature. The output of the bridge may be amplified and then delivered to a meter or some other device. In the operation of interest here, the output would be used to provide information for the microprocessor.

There are other nonlinear resistors that are used in much the same way as the thermistor. Here is a brief list of some of the more popular examples:

Photoresistors, which are sometimes called photoconductive devices, or, light-dependent resistors (LDRs) convert a level of light into a resistance value.

Voltage Dependent Resistors (or VDRs) convert a voltage change into a change in resistance.

144

Bolometers convert a small change in temperature to a change in resistance.

Photodiodes (also called light-activated diodes) conduct in a forward direction when exposed to light.

All of these light-sensing devices can be used in the same way as shown for the thermistor in Fig. 8-5.

Variable resistors can also be used as transducers. As a rule a certain amount of shaft rotation produces a resistance change that is used for sensing information. For the simple example shown in Fig. 8-6, the height of the fluid in the tank determines the setting of the variable resistor. That resistance, in turn, is used for information by the microprocessor.

The capacitance of a capacitor is determined by the area of the plates facing each other, the distance between those plates and the type of dielectric or insulating material between the plates. Any one of these three parameters may be varied in accordance with some quantity to be sensed. Examples are shown in Fig. 8-7.

In each application the capacitance is changed in some way, and, that change in capacitance affects the total impedance of a circuit.

The capacitor may be part of an oscillator circuit. In that case the change in capacity would change the output frequency of the oscillator. The frequency would be sensed or evaluated for measurement.

Another way to use capacitive transducers is to connect them in the leg of an ac bridge. This is also shown in Fig. 8-8. This bridge works like the one in Fig. 8-5 except the legs are impedances instead of resistances. They must be matched in order for a balance to occur.

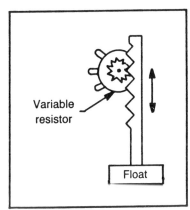

Fig. 8-6. The variable resistor is used as a transducer.

Variable resistor

Float

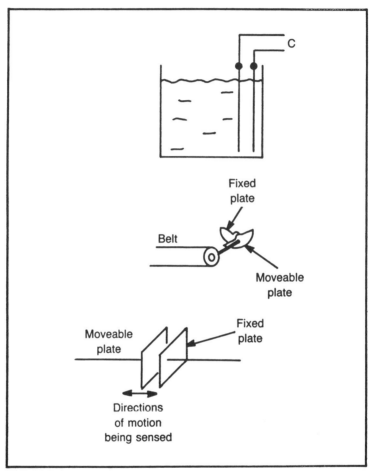

Fig. 8-7. Capacitor transducers work by changing the dielectric, area of plates facing each other, or distance between the plates.

If the impedance is changed (by the change in capacitance) then the bridge will be unbalanced by an amount that is directly related to the change in capacitance.

The inductance of an inductor depends upon the number of turns of wire and the core upon which the wire is wound. An example of an inductive transducer is shown in Fig. 8-9.

The typical inductive transducer changes the amount or type of core material. That, in turn, changes the inductance.

Inductances are used in ac bridges and in frequency-determining circuits. In that way they are very similar to capacitive transducers.

146

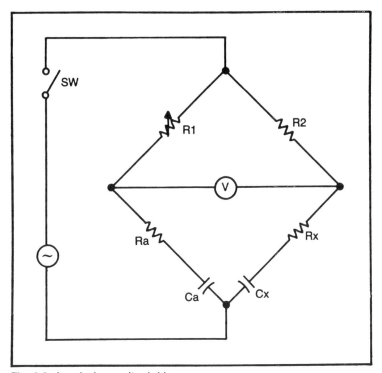

Fig. 8-8. A typical capacitor bridge.

Active Transducers

Basic textbooks in electronics usually list six different methods of generating a voltage. Theoretically, any of these six methods

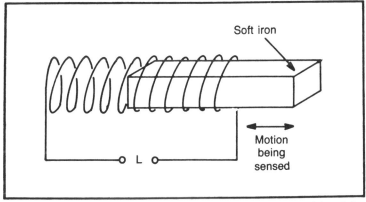

Fig. 8-9. The amount of inductance depends upon how far the soft iron is inserted into the coil.

could be used as a transducer but some of them are not ideally suited for the purpose.

- A voltage is created when two dissimilar insulating materials are rubbed together. This is the method of producing the voltage created when you stroke a cat's fur or when you walk across a carpet. It is sometimes called the friction method of generating a voltage.

    Commercially, very high voltages are created by this method in the form of electrostatic generators. These electrostatic generators can produce millions of volts and are used extensively for experimenting, measuring breakdown voltages and operating x-ray tubes. However, this method of generating a voltage is not normally used as a transducer for microprocessor input (friction).

- There are certain materials that will produce a voltage when a light is shined on them. This is called the photoelectric method of generating a voltage. It is used in transducers that convert a light energy input to an output voltage.

    Photoelectric transducers are used extensively in optical readers and other applications for microprocessor inputs.

- Certain crystalline materials will produce a voltage between their surfaces when a pressure is exerted upon them. They are called piezoelectric transducers. The amount of output voltage from these devices depends directly upon the amount of pressure created. Piezoelectric transducers are used very extensively to provide a microprocessor with information on the amount of pressure exerted at some point.

- When the junction of two dissimilar metals is inserted into a high-temperature region, a voltage is produced. This type of interface between dissimilar metals is referred to as a thermocouple or TE generator. (The TE stands for thermoelectric.) The output voltage is dependent upon the amount of temperature present and this is used as information input to microprocessors in systems where temperature is being sensed.

- Anytime two dissimilar metals are inserted in an acid or alkali solution, a voltage is generated. This is one of the earliest methods of producing a continuous voltage but it is not often used as a transducer input.

- Whenever a conductor is moved through a magnetic field there is always a voltage produced. The amount of voltage depends directly upon the speed at which the conductor is moved relative to the magnetic field. This is called the electromechanical method of producing a voltage. It is the principle upon which large generators produce electricity for cities. It is the principle upon which generators and alternators work. This principle can also be used as a transducer input. In the simple example of Fig. 8-10, acceleration causes the magnet to be thrust inside the coil. The greater the acceleration the more rapidly the magnet moves and the higher the amplitude of the impulse voltage produced.

SOME ADDITIONAL INTEGRATED CIRCUITS

The basic microprocessor system consists of the microprocessor, memory, and interfaces. Most microprocessor systems have more integrated circuits to simplify and extend their use. A few are given here.

A/D and D/A Converters

The output of most transducers is not suitable for direct input to the microprocessor for two reasons. One is that the voltage produced by these methods is very often too low. The other reason is that these are analog voltage values which must be converted so they are presented to the microprocessor in a digital form.

The way this is accomplished is to deliver the analog voltages to analog/digital (A/D) converters. They are available in integrated circuit form.

In the analog/digital converter the input is an analog voltage (such as might be obtained from the thermoelectric transducer) and the output is a digital signal that is related to the amplitude of the input signal.

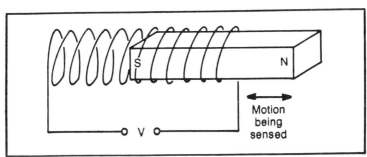

Fig. 8-10. Motion of the magnet induces a voltage in the coil.

As shown in Fig. 8-11, the signal from the transducer may be amplified before being presented to the A/D converter.

Digital/analog converters are also available. They take the output from the microprocessor (which is inherently a digital signal) and convert it into an analog signal such as audio or light intensity.

Some Uses of Monitors

The startup of a microprocessor system involves many steps. The registers and accumulators have to be cleared. The program counter has to be reset and a number of other functions must be taken care of before the microprocessor is ready to go.

When a microprocessor has been interrupted, it must store the contents of the counter, the registers and the accumulators into memory. After the interruption has been completed, all of the stored information must be returned from memory to its rightful place.

The procedure just described could be accomplished by having a person implement each step by using a keyboard. That would be a very tedious procedure because some microprocessors require many steps in the startup or restart procedure.

To get around this, the complete startup procedure is usually placed in a ROM. The procedure is automatically put into operation when the startup situation occurs. There may be a separate ROM for this called the *monitor*, or, this startup procedure may be incorporated into the ROM that is part of the basic microprocessor system.

The Assembler

The microprocessor receives its data and instructions on busses. Each line of the bus carries a digital number (1 or 0). Digital numbers are technically referred to as bits which stands for binary digits.

You *could* program the microprocessor with switches to set up these numbers. This was discussed earlier in this chapter. When

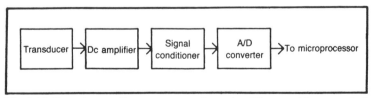

Fig. 8-11. Transducer voltages are normally analog. They must be converted to digital for microprocessor work.

you operate the microprocessor with individual combinations of ones and zeroes, you are using what is known as *machine language*.

It is easier to use mnemonics for programming. A computer program that automatically converts assembly language mnemonics into machine language is called an assembler. The mnemonics of assembly language are easy to remember and they are represented by hexidecimal codes. A few examples are given here:

ABA   Add Accumulators
AND   Logical And
CBA   Compare Accumulators
CLR   Clear
LDA   Load Accumulator A
SBA   Subtract Accumulators

The assembler is usually provided by the manufacturer in some form of read only memory. Higher languages, such as Basic, are converted to machine language by a program called compiler.

As in the case of the assembler program, compiler is provided by the manufacturer in the form of read only memory.

Phase-Locked Loops

A *phase-locked loop* is a closed-loop system with many applications. It is so important in microprocessor theory, because it is utilized with microprocessors that a review of the PLL system is necessary.

Some technicians will remember the use of the phase-locked loop circuit in the television horizontal oscillator of a TV receiver. Its purpose was to hold the oscillator frequency to an output value that is equal to the horizontal sync pulse input.

Another earlier application of the phase-locked loop was in single-conversion receivers.

It was not until the phase-locked loop was put on an integrated circuit package that it became very popular.

Figure 8-12 shows the basic loop circuitry. In this system there is a *phase detector* that compares the phase of an input frequency from an external source (such as a crystal oscillator) with a second input frequency from the voltage-controlled oscillator (VCO). These two inputs are compared in phase. If there is a shift in phase it indicates that an oscillator is drifting off frequency.

Assume that there is a phase difference. If input *fa* is from a

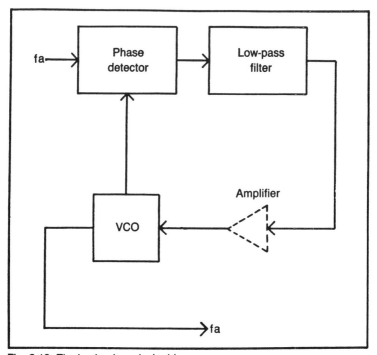

Fig. 8-12. The basic phase-locked loop.

crystal oscillator, then it can be assumed that the VCO is the one that is drifting.

The correction voltage output of the phase detector is delivered to a low-pass filter that removes fa, the VCO frequency, and their sum. Only the difference frequency—which is a dc voltage proportional to the phase difference between the two input signals—is delivered through the low-pass filter to the dc amplifier.

The dc amplifier is shown in broken lines because it is not used in every phase-locked loop. When it is included, it is used for the purpose of increasing the sensitivity of the loop system.

The dc voltage at the output of the low-pass filter is used with or without amplification to set the frequency of the voltage controlled oscillator. Generally, the oscillator frequency control device is a voltage-dependent capacitor called a *varactor diode*. This varactor forms part of the voltage controlled oscillator frequency-determining circuitry.

Having defined the various parts of the loop, let's go around the loop again. Assume that the VCO has drifted off frequency so that it is higher in frequency than fa. This will cause a dc correction voltage out of the phase comparator. It is filtered (and sometimes amplified) and then delivered back to the frequency determining network of the VCO. The feedback voltage will always be such that it will bring the oscillator back into phase (and frequency) with input *fa*.

Since the VCO output frequency is phase-locked to *fa* it follows that the output frequency of the VCO is exactly equal to *fa*.

The basic phase-locked loop circuit can be used for an FM detector. Remember that the frequency of an FM signal varies in accordance with some audio input at the transmitter. The audio, then, causes the frequency of the carrier to move up and down with changes in audio amplitude.

In the phase-locked loop system of Fig. 8-13 the FM signal is delivered to the phase detector loop. As the frequency varies up and down (in accordance with the audio), the output correction voltage also varies up and down from instant to instant. The result is that the output from the low-pass filter is the original modulating audio signal.

This is not a microprocessor application but it is included here to show a practical application of the phase-locked loop.

Since the output frequency from the VCO in Fig. 8-12 is the same as the input frequency, not much has been accomplished except that the VCO output is crystal stabilized by the crystal control frequency input.

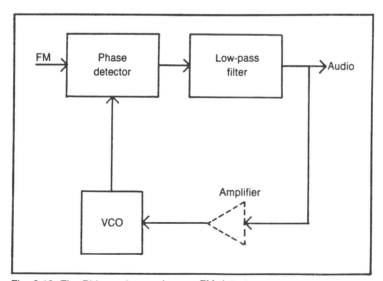

Fig. 8-13. The PLL can be used as an FM detector.

153

Consider now the phase-locked loop in Fig. 8-14. A divide by N (÷N) integrated circuit has been added between the VCO and the phase detector (ϕ).

You will remember that in order for a phase-lock to occur the two inputs into the phase detector (*fa* and *fb*) must be equal. The only way this can be accomplished is that the VCO frequency must be N times the input frequency. That is necessary so that when you divide the VCO frequency by N it is equal to *fa*.

A simple example will show how this works. Assume that the input frequency is 100 kilohertz. That means that *fa* is 100 kilohertz, and *fb* must also be 100 kilohertz if a phase-lock is to be accomplished. If *fb* is 100 kilohertz after being divided by 2, it follows that the VCO frequency must be 200 kilohertz. So, the output frequency is *fb* or 200 kilohertz.

Now you see that the output frequency of the VCO is twice the crystal control frequency. It is, in itself, highly stable because it is locked to a crystal oscillator frequency. If you wanted an output

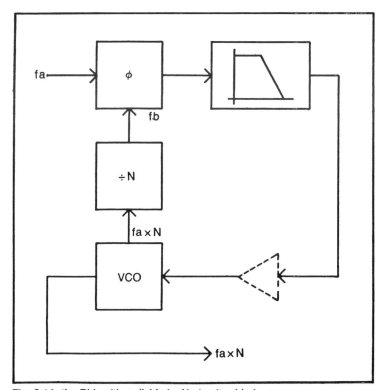

Fig. 8-14. the PLL with a divide-by-N circuit added.

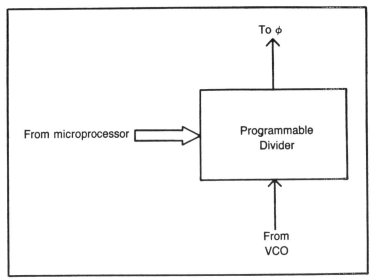

To φ

From microprocessor ⟹ Programmable Divider

From VCO

Fig. 8-15. A programmable divider is used as a divide-by-N circuit.

VCO of 300 kilohertz it would be a simple matter to set the divide-by-N circuit equal to divide by 3.

In order to make this circuit very useful, a programmable divider is usually used for the divide-by-N circuit. This type of device is illustrated in Fig. 8-15. The VCO output goes to the programmable divider and the output of that circuit goes to the phase comparator.

A microprocessor delivers a coded signal that determines the number used for a divisor. In a previous example the microprocessor would deliver a coded input to the programmable divider to select a divide by 2 mode of operation. Note that when this programmable divider is used F for divide by N in the circuit of Fig. 8-14 the microprocessor actually determines the output frequency from the VCO by the code that it delivers to the programmable divider.

Multiplexers and Demultiplexers

Microprocessors are identified by the number of lines in their data bus. For example, an 8-bit microprocessor has an eight line data bus.

Four-bit microprocessors are popular in dedicated applications. Examples are the microprocessor used in some appliances and automotive applications. The obvious disadvantage of a four-bit microprocessor is that there is a limit to the number of inputs it can handle.

One way to get around this is to use a multiplexer. This type of integrated circuit permits more than one input to be reduced to a single output. This is accomplished by switching the inputs to the output line one at a time.

Demultiplexers work in an opposite way. In the case of the four-bit microprocessor they take the outputs on four data lines and reduce them to a single line output.

APPLICATIONS

As you might expect from what has been said in this book, applications of microprocessors involve uses of memory in some way. A few examples are given here.

An FM Tuner

In this application the microprocessor makes it possible to preselect a number of the customer's favorite FM stations using a single push-button for tuning. A brief review of the FM tuner is necessary for a clear understanding of how this system works.

The FM broadcast band is spread between 88 and 108 megahertz. There are 99 carriers in the FM band that are spaced between 88.1 and 107.9 megahertz. In other words, there are 99 individual stations possible in the 88 to 108 megahertz range.

The intermediate frequency of FM receivers has been standardized at 10.7 megahertz. As shown in the block diagram of Fig. 8-16, all of the stations are picked up by the antenna and delivered to the rf amplifier. This amplifier is tuned to select the desired station and the tuned signal is delivered to a mixer stage.

A local oscillator delivers a constant frequency signal to the mixer and a difference (or, intermediate) frequency of 10.7 megahertz is obtained. This frequency is selected at the mixer output.

In order to maintain the 10.7 megahertz i-f frequency, it is necessary that the local oscillator be tuned for each station selection. In the application to be discussed the local oscillator frequency is selected by the microprocessor. Remember, this frequency is mixed with the desired station to get the i-f frequency.

Note the equation in Fig. 8-16. The rf frequency and the i-f frequency when added together equal the required oscillator frequency.

Figure 8-17 shows the basic system for producing that oscilla-

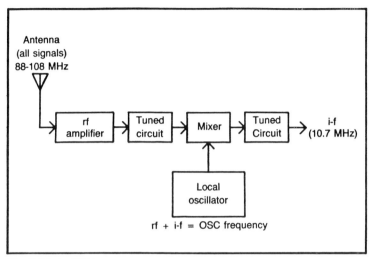

Fig. 8-16. An FM tuner.

tor frequency. Observe that there is a phase-locked loop in this system. The phase comparator is marked with the symbol phi (ϕ). That phase comparator compares an input from a crystal oscillator with the input from the VCO.

In the case of the crystal oscillator the 200 kilohertz steady frequency is divided by 8 to produce a 25 kilohertz input to the phase comparator. That input is compared with a 25 kilohertz signal from the loop.

The output of the phase comparator goes to a low-pass filter. It is depicted with a typical low-pass filter response curve.

There is no amplifier in this phase-locked loop. The output of the filter goes directly to the VCO. This output is a dc correction voltage from the phase comparator that keeps the VCO on the desired frequency.

The output of the VCO is divided by a prescaler and then by a divide-by-N circuit. After the two divisions, the input to the phase comparator is 25 kilohertz.

Assume that the customer has selected as one of his push-button choices the FM station broadcasting at 104.1 megahertz. In that case the microprocessor will be set up to divide by 574 as shown in Fig. 8-18. Note that the oscillator frequency must be 114.8 megahertz in this case. That is obtained from the equation

$$
\begin{aligned}
\text{rf} + \text{i-f} &= \text{oscillator frequency} \\
114 + 10.7 &= 114.8 \text{ megahertz}
\end{aligned}
$$

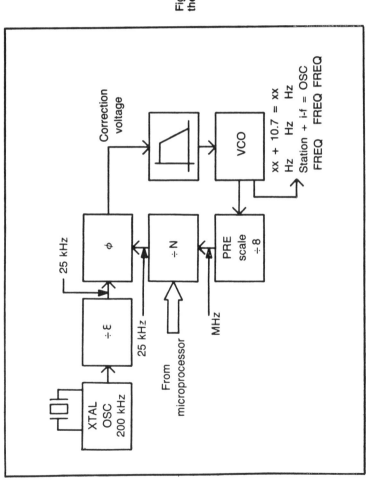

Fig. 8-17. Using the PLL to obtain the local oscillator frequency.

158

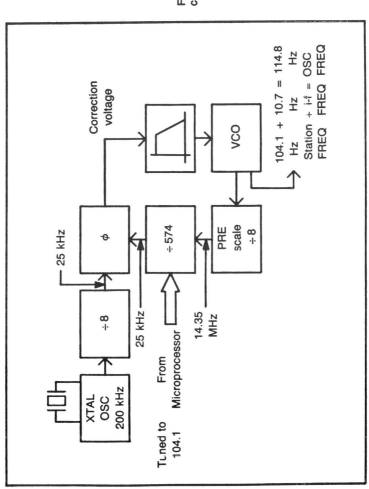

Fig. 8-18. A specific example of a local oscillator frequency.

When the 114.8 megahertz is divided by 8, there is a 14.35 megahertz signal. That signal, when divided by 574 in the divide-by-N circuit, will give a 25 kilohertz input to the phase comparator.

What is the job of the microprocessor in this case? It must produce a coded input to the divide-by-N circuit in order to obtain the correct oscillator frequency. The microprocessor is used in a *single-stepping mode* so that each time the push-button is operated the microprocessor is changed to the next preselected input from the electrically erasable read-only memory. The electrically erasable read-only memory is used here so that the customer or technician can, in the field, preselect the stations desired.

The microprocessor goes to the EEROM and brings the coded signal into an accumulator. Then, that coded signal is delivered through an interface to the divide-by-N circuit in Fig. 8-18.

The divide-by-N circuit that receives the microprocessor coded signal is a *programmable divider*. Its operation is illustrated in Fig. 8-15. The number that the input signal is divided by depends upon a binary code delivered on a four line bus. This permits 2 to the 4th power, or, a total of 16 selections.

To summarize, it is the job of the microprocessor to select a code from memory and deliver it to the programmable divider. That in turn sets up the frequency of the VCO for use in the FM tuner.

In another application the microprocessor and phase-lock loop can be used to accurately determine the speed of a stepping motor (or *synchronous motor*). See Fig. 8-19. A synchronous motor has a speed that is directly dependent upon its input frequency. An example of its application is an electric clock designed to operate from the power line. The power line frequency is a very accurate 60 hertz which accurately sets the speed of rotation of the synchronous motor.

The shaft of a stepping motor turns an exact number of degrees for each impulse on its power input. If the stepping motor shaft turns 36 degrees with each impulse, then 10 pulses per second will result in one revolution per second.

The voltage controlled oscillator in this application produces an output frequency to a prescaler. The output of the prescaler is divided by N to obtain a signal that is compared with the output of the divided crystal oscillator frequency. If the two are equal in frequency the low-pass filter delivers no correction voltage to the VCO. If the VCO begins to drift off frequency the closed loop is such that it will bring the oscillator back to the desired frequency.

The frequency of the VCO in this system sets the speed of the

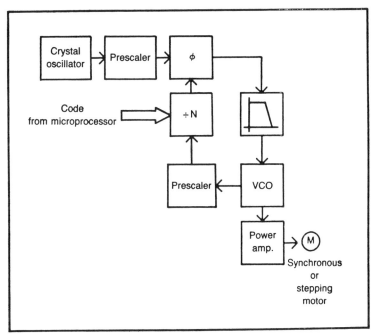

Fig. 8-19. This simplified drawing shows one of the many ways that a microprocessor can be used to control motor speed.

motor. The microprocessor in this case, as in the previous case, delivers a code to the divide-by-N circuit. In reality that is what sets the VCO frequency and the motor speed. The desired codes for divide by N are stored in the microprocessor memory system. A keyboard can be used to set the desired motor speed.

Figure 8-20 shows how the microprocessor can be used as a burglar alarm. The switches are actually located at doors and/or windows. As shown by the inset, these switches can be in either a logic 1 or logic 0 position, corresponding to whether the door is open or closed. As a rule a logic 1 represents an open door.

The microprocessor would be of little use here if the application is such that all the doors and windows are to be open during the day and closed at night. However, in this application, positions A and D are normally closed 24 hours a day while B and C are normally open. So, during the daytime the input to the PIA is 0110. This input is delivered by the PIA to the microprocessor. The microprocessor compares it with a 0110 brought from memory.

In a typical application the two binary numbers are subtracted and there should be a 0 output on the alarm line. If any of the doors

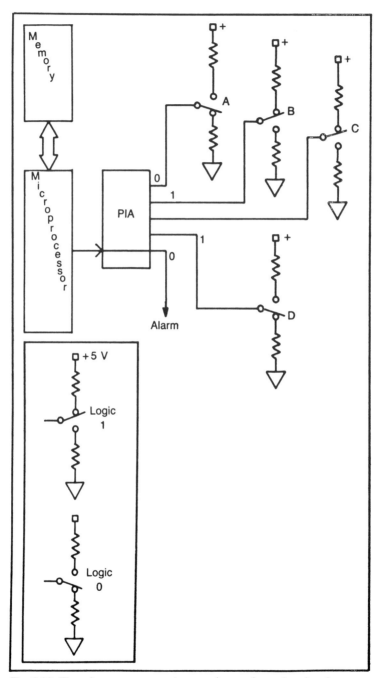

Fig. 8-20. The microprocessor system can be used as a burglar alarm.

or windows changes position, they will no longer subtract to 0 when matched with the number in memory.

If there is a change in the operation it is a simple matter to change the number in memory. Suppose, for example, that A is to be open for a period of time. That would give an input of 1110. The memory would be changed to 1110 so that when they are subtracted there would be a 0 output on the alarm line. So, this microprocessor operated burglar alarm is more flexible than one that is done with hard wired logic.

A disadvantage is that only four inputs are permitted as shown in the illustration. Remember, however, that multiplexers can be used at each of the four positions. A four-input multiplexer, for example, would give a total of 16 lines or 16 positions that could be continuously monitored.

This is not the only way that microprocessors are used in burglar alarms. The PIA is designed in such a way that it can monitor any single line input. In a procedure called poling the microprocessor can sequence the inputs one at a time. In other words, it checks the doors and windows one at a time to determine if the normal condition exists. In that method if there is a switch in the wrong position the microprocessor can produce a signal telling exactly which sense input is in the wrong switch position.

The applications we have given are primarily used to show that the microprocessor utilizes memory to perform various functions.

The microprocessor in this application may be used for many other duties in the plant. So, it only works a very small part of its total time as a burglar alarm. When a single microprocessor is used for a wide variety of applications, it is called *time sharing*. Even though the simple burglar alarm system of Fig. 8-18 could be made with hard wired logic, it would not be as useful as using the already existent microprocessor.

PROGRAMMED REVIEW

(See the instructions for using this section in Chapter 1.)

Block 1

Switches are not normally used for direct input of data to a microprocessor because

(a) the contacts bounce. *Go to Block 8.*
(b) there is very little distinction between logic 1 and logic 0. *Go to Block 12.*

Block 2

The correct answer to the question in Block 33 is choice (a). An amplifier is only used when it is necessary to increase the sensitivity of the loop system.

Here is your next question: The reference frequency into the phase detector of a phase-locked loop is 50 kHz. The divide-by-N circuit is set to divide the VCO output frequency by 5. The VCO output frequency must be

(a) 10 kilohertz. *Go to Block 21.*
(b) 250 kilohertz. *Go to Block 25.*

Block 3

The correct answer to the question in Block 29 is choice (b). Although capacitive and inductive transistors are used in ac bridges, they have a common purpose with dc bridge circuits like those used with thermistors. The bridge is unbalanced by changes in transducer input.

Here is your next question: Which of the following is used to convert a passive transducer input to a signal that can be used by a microprocessor?

(a) An A/D converter *Go to Block 9.*
(b) A D/A converter *Go to Block 18.*

Block 4

Your answer to the question in Block 26 is not correct. Read the question again, then *go to Block 11.*

Block 5

The correct answer to the question in Block 24 is choice (b). Using the equation in Fig. 8-14,

$$rf + i\text{-}f = \text{oscillator frequency}$$
$$88.7 + 10.7 = 99.4 \text{ megahertz}$$

Here is your next question: Which of the following motors has a speed that depends upon the input frequency?

(a) Synchronous motor *Go to Block 19.*

164

(b) Stepping motor. *Go to Block 31.*

(c) (Both choices are correct.) *Go to Block 27.*

Block 6

Your answer to the question in Block 25 is not correct. Read the question again, then *go to Block 34.*

Block 7

The answer to the question in Block 27 is RAM. By using the RAM it is an easy matter to determine which doors are protected by the alarm system.

Here is your next question: Which type of circuit combines more than one input into a single output? (multiplexer or demultiplexer) *Go to Block 36.*

Block 8

The correct answer to the question in Block 1 is choice (a). Switches make excellent binary indicators. Their disadvantages are slow speed and inconvenience.

Here is your next question: To overcome one of the difficulties that occurs when switches are used to program microprocessors, use a circuit called a

(a) debouncer. *Go to Block 14.*

(b) one-zero enhancer. *Go to Block 20.*

Block 9

The correct answer to the question in Block 3 is choice (a). An analog signal from either a passive or active transducer must be converted into a digital circuit that can be used by a microprocessor. That is accomplished by using an A/D converter.

Here is your next question: Which of the following is used to start—or restart—a microprocessor?

(a) Monitor. *Go to Block 26.*

(b) Starter switch. *Go to Block 30.*

Block 10

Your answer to the question in Block 34 is not correct. Read the question again, then *go to Block 24.*

Block 11

The correct answer to the question in Block 26 is choice (b). This is the procedure for interrupting a microprocessor.

Here is your next question: A computer program that converts hexadecimal codes—in the form of mnemonics—into machine language is called

(a) a programmer. *Go to Block 16.*
(b) an assembler. *Go to Block 22.*

Block 12

Your answer to the question in Block 1 is not correct. Read the question again, then *go to Block 8.*

Block 13

Your answer to the question in Block 33 is not correct. Read the question again, then *go to Block 2.*

Block 14

The correct answer to the question in Block 8 is choice (a). Debouncers are usually made with NANDs, NORs, or INVERTERS.

Here is your next question: Which of the following is an example of a passive transducer?

(a) Piezoelectric crystal. *Go to Block 23.*
(b) Thermistor. *Go to Block 29.*

Block 15

Your answer to the question in Block 24 is not correct. Read the question again, then *go to Block 5.*

Block 16

Your answer to the question in Block 11 is not correct. Read the question again, then *go to Block 22.*

Block 17

Your answer to the question in Block 34 is not correct. Read the question again, then *go to Block 24.*

Block 18

Your answer to the question in Block 3 is not correct. Read the question again, then *go to Block 9.*

Block 19

Your answer to the question in Block 5 is not correct. Read the question again, then *go to Block 27.*

Block 20

Your answer to the question in Block 8 is not correct. Read the question again, then *go to Block 14.*

Block 21

Your answer to the question in Block 2 is not correct. Read the question again, then *go to Block 25.*

Block 22

The correct answer to the question in Block 11 is choice (b). The assembler may be stored in the ROM, or, it may be a separate integrated circuit.

Here is your next question: A high-level language is converted to machine language by a

(a) programmer. *Go to Block 28.*
(b) compiler. *Go to Block 33.*

Block 23

Your answer to the question in Block 14 is not correct. Read the question again, then *go to Block 29.*

Block 24

The correct answer to the question in Block 34 is choice (c). The code comes from memory. A non-volatile memory is needed, so the memory must be a form of ROM. An EEROM is usually used because it is non-volatile, but its contents can be readily changed in the field.

Here is your next question: An FM station with a frequency

of 88.7 megahertz is tuned by a customer. The tuner oscillator frequency must be

(a) 88.7 megahertz. *Go to Block 32.*
(b) 99.4 megahertz. *Go to Block 5.*
(c) (Neither choice is correct.) *Go to Block 15.*

Block 25

The correct answer to the question in Block 2 is choice (b). Dividing the VCO frequency by 5 must result in a 50 kilohertz signal into the phase comparator. So, 250 kHz is divided by 5 gives 50 kHz.

Here is your next question: The frequency of a VCO is determined by a dc voltage delivered to a

(a) varactor diode. *Go to Block 34.*
(b) tantalum capacitor. *Go to Block 6.*

Block 26

The correct answer to the question in Block 9 is choice (a). The monitor is in the form of a read-only memory. It may be a separate integrated circuit on the microprocessor board.

Here is your next question: When a microprocessor is interrupted in the performance of its duties

(a) the information in the program counter, registers and accumulators is lost. The program must be restarted. *Go to Block 4.*
(b) the information in the program counter, registers and accumulators is stored in memory until the interruption is over. Then, all of the information is put back where it belongs. *Go to Block 11.*

Block 27

The correct answer to the question in Block 5 is choice (c). Other types of motors (besides the stepping and synchronous types) can be controlled by microprocessors.

Here is your next question: To make the burglar alarm system of Fig. 8-18 more versatile, the code for matching the input from the PIA is stored in (*RAM or ROM*). *Go to Block 7.*

Block 28

Your answer to the question in Block 22 is not correct. Read the question again, then *go to Block 33*.

Block 29

The correct answer to the question in Block 14 is choice (b). A thermistor is a non-linear resistor. Its resistance value changes over a wide range of ohms for a relatively small change in temperature.

Here is your next question: ac bridges for capacitive and inductive transducers use

(a) resistances instead of impedances in the legs of the bridge circuits. *Go to Block 35*.
(b) impedances instead of resistances in the legs of the bridge circuits. *Go to Block 3*.

Block 30

Your answer to the question in Block 9 is not correct. Read the question again, then *go to Block 26*.

Block 31

Your answer to the question in Block 5 is not correct. Read the question again, then *go to Block 27*.

Block 32

Your answer to the question in Block 24 is not correct. Read the question again, then *go to Block 5*.

Block 33

The correct answer to the question in Block 22 is choice (b). Compilers are used extensively in computers.

Here is your next question: Which of the following is not always part of a phase-locked loop?

(a) Dc Amplifier *Go to Block 2*.
(b) VCO *Go to Block 13*.

Block 34

The correct answer to the question in Block 25 is choice (a). Varactor diodes are sometimes called voltage-variable capacitors. The amount of reverse voltage across the diode determines its junction capacitance. The higher the reverse voltage value the lower the capacitance value.

Here is your next question: The microprocessor delivers a code to the programmable divider. It gets this code from

(a) the monitor. *Go to Block 10.*
(b) the assembler. *Go to Block 17.*
(c) (Neither choice is correct.) *Go to Block 24.*

Block 35

Your answer to the question in Block 29 is not correct. Read the question again, then *go to Block 3.*

Block 36

The correct answer to the question in Block 7 is Multiplexer. You have now completed the Programmed Review section.

EXPERIMENTS

In the experiments for this chapter you will construct and operate some multiplex circuitry.

A Typical Multiplexer

Figure 8-21 shows the manufacturer's detailed specification for a four-line to one-line multiplexer. The 74153 is very easy to work with.

Connect the I(a) inputs to logic 1. Then, with the proper input logic levels to the Ea and S1 and S0 inputs, step through the four "A" positions. This is done by setting NOT Ea to logic 0 and sequencing S0 and S1 through a binary count. It is helpful to use LEDs at the S0 and S1 inputs so you can keep track of the binary count as you step through the various multiplexed inputs.

The output at pin 7 should be logic 1 for each count. Next, wire the I(a) inputs to logic 0 and repeat the procedure. In that case the output should be at 0 at all times.

Rewire the I(a) inputs for 0 1 1 0 and single step through the binary count. Note that the output delivers the logic inputs—one at a time.

MULTIPLEXERS

Dual 4-Line To 1-Line Multiplexer

- **Non-inverting outputs**
- **Separate Enable for each section**
- **Common Select inputs**
- **See '253 for 3-State version**

TYPE	TYPICAL PROPAGATION DELAY	TYPICAL SUPPLY CURRENT (Total)
74153	18ns	36mA
74LS153	18ns	6.2mA
74S153	9ns	45mA

ORDERING CODE

PACKAGES	COMMERCIAL RANGES $V_{CC} = 5V \pm 5\%; T_A = 0°C$ to $+70°C$	MILITARY RANGES $V_{CC} = 5V \pm 10\%; T_A = -55°C$ to $+125°C$	
Plastic DIP	N74153N • N74LS153N N74S153N		
Plastic SO	N74LS153D • N74S153D		
Ceramic DIP		S54153F • S54LS153F S54S153F	
Flatpack		S54153W • S54LS153W S54S153W	
LLCC		S54S153G • S54LS153G	

DESCRIPTION

The '153 is a dual 4-input multiplexer that can select 2 bits of data from up to four sources under control of the common Select inputs (S_0, S_1). The two 4-input multiplexer circuits have individual active LOW Enables ($\bar{E}_a$, $\bar{E}_b$) which can be used to strobe the outputs independently. Outputs (Y_a, Y_b) are forced LOW when the corresponding Enables ($\bar{E}_a$, $\bar{E}_b$) are HIGH.

The device is the logical implementation of a 2-pole, 4-position switch, where the position of the switch is determined by the logic levels supplied to the two Select inputs. The logic equations for the outputs are shown below.

$$Y_a = \bar{E}_a \cdot (I_{0a} \cdot \bar{S}_1 \cdot \bar{S}_0 + I_{1a} \cdot \bar{S}_1 \cdot S_0 + I_{2a} \cdot S_1 \cdot \bar{S}_0 + I_{3a} \cdot S_1 \cdot S_0)$$

$$Y_b = \bar{E}_b \cdot (I_{0b} \cdot \bar{S}_1 \cdot \bar{S}_0 + I_{1b} \cdot \bar{S}_1 \cdot S_0 + I_{2b} \cdot S_1 \cdot \bar{S}_0 + I_{3b} \cdot S_1 \cdot S_0)$$

The '153 can be used to move data to a common output bus from a group of registers. The state of the Select inputs would

INPUT AND OUTPUT LOADING AND FAN-OUT TABLE

PINS	DESCRIPTION	5474	54S74	5474LS
All	Inputs	1ul	1Sul	1LSul
All	Outputs	10ul	10Sul	10LSul

NOTE
Where a 54/74 unit load (ul) is understood to be 40μA I_{IH} and $-1.6mA$ I_{IL}, and a 54/74S unit load (Sul) is 50μA I_{IH} and $-2.0mA$ I_{IL}, and a 54/74LS unit load (LSul) is 20μA I_{IH} and $-0.4mA$ I_{IL}.

determine the particular register from which the data came. An alternative application is as a function generator. The device can generate two functions or three variables. This is useful for implementing highly irregular random logic.

Fig. B-21. Manufacturer's specification for the 74153 multiplexer. (courtesy of Signetics, Inc.) Continued to page 173.)

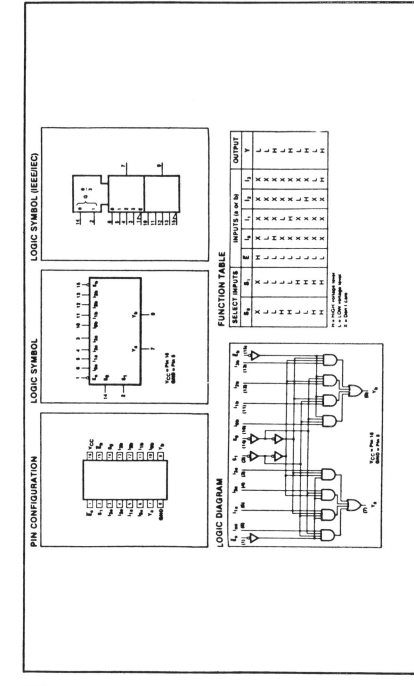

PIN CONFIGURATION

LOGIC SYMBOL

LOGIC SYMBOL (IEEE/IEC)

FUNCTION TABLE

SELECT INPUTS			INPUTS (a or b)				OUTPUT
S_0	S_1	E	I_0	I_1	I_2	I_3	Y
X	X	H	X	X	X	X	L
L	L	L	L	X	X	X	L
L	L	L	H	X	X	X	H
H	L	L	X	L	X	X	L
L	H	L	X	H	X	X	H
L	H	L	X	X	L	X	L
H	H	L	X	X	H	X	H
H	H	L	X	X	X	L	L
H	H	L	X	X	X	H	H

H = HIGH voltage level
L = LOW voltage level
X = Don't care

LOGIC DIAGRAM

172

ABSOLUTE MAXIMUM RATINGS (Over operating free-air temperature range unless otherwise noted.)

	PARAMETER	54	54LS	54S	74	74LS	74S	UNIT
V_{CC}	Supply voltage	7.0	7.0	7.0	7.0	7.0	7.0	V
V_{IN}	Input voltage	-0.5 to +5.5	-0.5 to +7.0	-0.5 to +5.5	-0.5 to +5.5	-0.5 to +7.0	-0.5 to +5.5	V
I_{IN}	Input current	-30 to +5	-30 to +1	-30 to +5	-30 to +5	-30 to +1	-30 to +5	mA
V_{OUT}	Voltage applied to output in HIGH output state	-0.5 to +V_{CC}	-0.5 to +V_{CC}	-0.5 to +V_{CC}	-0.5 to +V_{CC}	-0.5 to +V_{CC}	-0.5 to +V_{CC}	V
T_A	Operating free-air temperature range	-55 to +125	-55 to +125	-55 to +125	0 to 70	0 to 70	0 to 70	°C

RECOMMENDED OPERATING CONDITIONS

	PARAMETER		5474 Min	5474 Nom	5474 Max	5474LS Min	5474LS Nom	5474LS Max	54/74S Min	54/74S Nom	54/74S Max	UNIT
V_{CC}	Supply voltage	Mil	4.5	5.0	5.5	4.5	5.0	5.5	4.5	5.0	5.5	V
		Com'l	4.75	5.0	5.25	4.75	5.0	5.25	4.75	5.0	5.25	V
V_{IH}	HIGH-level input voltage	Mil	2.0			2.0			2.0			V
V_{IL}	LOW-level input voltage	Mil			+0.8			+0.7			+0.8	V
		Com'l			+0.8			+0.8			+0.8	V
I_{IK}	Input clamp current	Mil			-12			-18			-18	mA
I_{OH}	HIGH-level output current	Mil			-800			-400			-1000	µA
I_{OL}	LOW-level output current	Mil			16			4			20	mA
		Com'l			16			8			20	mA
T_A	Operating free-air temperature	Mil	-55		+125	-55		+125	-56		+125	°C
		Com'l	0		70	0		70	0		70	°C

NOTE
V_{IL} = +0.7V MAX for 54S at T_A = +125°C only.

Demultiplexer

Figure 8-22 shows the manufacturer's specifications for the 74155 decoder/demultiplexer. In order to use this as a demultiplexer the input information must be delivered to one of the enables and the other tied to logic 0 or 1—depending upon whether you want the input to be *inverted* or *not inverted*. For example, if you do not want the input to be inverted, deliver the input signal to pin 1 and tie pin 2 to logic 0.

It is necessary to sequence through the outputs by delivering a binary count to the address inputs (pins 3 and 13).

Use LEDs for the multiplex outputs at pins 4, 5, 6 and 7. Then, tie the input (pin 1) to a logic 1. This means you are going to deliver a logic 1 to each of the outputs (pins 4 through 7) one at a time.

Single step the address inputs through the binary count and note that the outputs sequence as indicated by the LEDs. What you are doing is delivering a logic input to four logic lines.

Multiplex/Demultiplex Configuration

Use a multiplexer to convert four lines of information to a single line. Then, deliver that single line to a demultiplexer. The four outputs from the demultiplexer should have exactly the same logic configuration as the four inputs of the multiplexer.

You will have to be careful in this experiment to make sure that the binary counts of the address inputs for both integrated circuits have the same identical numbers. In other words, both counts must be at 00 at the same time, 01 at the same time, etc.

SELF TEST

1. Which of the following is a passive transducer?

   (a)   Variable resistor
   (b)   Variable capacitor
   (c)   Fixed inductor
   (d)   (All are passive transducers.)

2. If an FM receiver is tuned to 101.3 megahertz, the local oscillator frequency will be

   (a)   101.3 megahertz

174

DECODERS/DEMULTIPLEXERS

54/74155, LS155

Dual 2-Line To 4-Line Decoder/Demultiplexer

- Common Address inputs
- True or complement data demultiplexing
- Dual 1-of-4 or 1-of-8 decoding
- Function generator applications

DESCRIPTION

The '155 is a Dual 1-of-4 Decoder/Demultiplexer with common Address inputs and separate gated Enable inputs. Each decoder section, when enabled, will accept the binary weighted Address input (A_0, A_1) and provide four mutually exclusive active-LOW outputs ($\bar{0}$-$\bar{3}$). When the enable requirements of each decoder are not met, all outputs of that decoder are HIGH.

Both decoder sections have a 2-input enable gate. For decoder "a" the enable gate requires one active-HIGH input and one active-LOW input (E_a, $\bar{E}_a$). Decoder "a" can accept either true or complemented data in demultiplexing applications, by using the $\bar{E}_a$ or E_a inputs respectively. The decoder "b" enable gate requires two active-LOW inputs ($\bar{E}_b$, $\bar{E}_b$). The device can be used as a 1-of-8 decoder/demultiplexer by tying E_a to $\bar{E}_b$ and relabeling the common connection address as (A_2); forming the common enable by connecting the remaining $\bar{E}_b$ and $\bar{E}_a$.

TYPE	TYPICAL PROPAGATION DELAY	TYPICAL SUPPLY CURRENT (Total)
74155	18ns	25mA
74LS155	17ns	6.1mA

ORDERING CODE

PACKAGES	COMMERCIAL RANGES $V_{CC} = 5V \pm 5\%; T_A = 0°C$ to $+70°C$	MILITARY RANGES $V_{CC} = 5V \pm 10\%; T_A = -55°C$ to $+125°C$
Plastic DIP	N74155N • N74LS155N	
Plastic SO	N74LS155D	

INPUT AND OUTPUT LOADING AND FAN-OUT TABLE

PINS	DESCRIPTION	54/74	54/74LS
All	Inputs	1ul	1LSul
All	Outputs	10ul	10LSul

NOTE
Where a 54/74 unit load (ul) is understood to be 40µA I_{IH} and −1.6mA I_{IL}, and a 54/74LS unit load (LSul) is 20µA I_{IH} and −0.4mA I_{IL}.

Fig. 8-22. Manufacturer's specification for the 74155 demultiplexer. (courtesy of Signetics, Inc.) (Continued from page 177.)

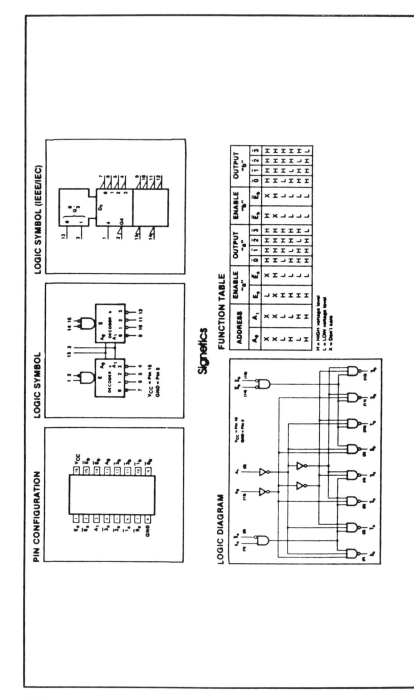

Signetics

PIN CONFIGURATION

LOGIC SYMBOL

LOGIC SYMBOL (IEEE/IEC)

FUNCTION TABLE

LOGIC DIAGRAM

ABSOLUTE MAXIMUM RATINGS (Over operating free-air temperature range unless otherwise noted.)

PARAMETER		54	54LS	74	74LS	UNIT
V_{CC}	Supply voltage	7.0	7.0	7.0	7.0	V
V_{IN}	Input voltage	−0.5 to +5.5	−0.5 to +7.0	−0.5 to +5.5	−0.5 to +7.0	V
I_{IN}	Input current	−30 to +5	−30 to +1	−30 to +5	−30 to +1	mA
V_{OUT}	Voltage applied to output in HIGH output state	−0.5 to +V_{CC}	−0.5 to +V_{CC}	−0.5 to +V_{CC}	−0.5 to +V_{CC}	V
T_A	Operating free-air temperature range	−55 to +125		0 to 70		°C

RECOMMENDED OPERATING CONDITIONS

PARAMETER		5474 Min	5474 Nom	5474 Max	5474LS Min	5474LS Nom	5474LS Max	UNIT
V_{CC} Supply voltage	Mil	4.5	5.0	5.5	4.5	5.0	5.5	V
	Com'l	4.75	5.0	5.25	4.75	5.0	5.25	V
V_{IH} HIGH-level input voltage	Mil	2.0			2.0			V
	Com'l							V
V_{IL} LOW-level input voltage	Mil			+0.8			+0.7	V
	Com'l			+0.8			+0.8	V
I_{IK} Input clamp current	Mil			−12			−18	mA
I_{OH} HIGH-level output current	Mil			−800			−400	µA
	Com'l							
I_{OL} LOW-level output current	Mil			16			4	mA
	Com'l			16			8	mA
T_A Operating free-air temperature	Mil	−55		+125	−56		+125	°C
	Com'l	0		70	0		70	°C

(b)   10.7 megahertz
(c)   112 megahertz
(d)   (None of these choices is correct.)

3. Which of the following types of motors will run at a speed that is dependent upon the frequency of the power?

(a)   Shunt wound dc
(b)   Synchronous
(c)   Three phase induction under heavy load
(d)   Simple Compound-Wound dc

4. When switches are used as microprocessor inputs they are connected through a circuit that eliminates the effect of

(a)   contact bounce.
(b)   high contact current.
(c)   high contact voltage.
(d)   contact resistance.

5. Keyboard switches are connected to the microprocessor through

(a)   a direct memory access integrated circuit.
(b)   an encoder.
(c)   a decoder.
(d)   an ACIA.

6. Which of the following is not normally used as a transducer?

(a)   Static electricity
(b)   Piezoelectricity
(c)   Photoelectricity
(d)   Thermoelectricity

7. Before transducer voltages can be delivered to a microprocessor they must go through

(a)   a frequency synthesizer.
(b)   a compiler.
(c)   an A/D converter.
(d)   a D/A converter.

8. Mnemonics are directly related to hexidecimal codes. These hexidecimal codes are converted to machine language in

(a) a synchronizer.
(b) an A/D converter.
(c) an ACIA.
(d) an assembler.

9. Which of the following is not always included in a phase-locked loop?

(a) Phase detector
(b) Low-pass filter
(c) dc amplifier
(d) VCO

10. Which of the following is used for a divide-by-N circuit in a microprocessor-controlled phase-locked loop?

(a) ALU
(b) Programmable Divider
(c) Controller
(d) Shift Register

Answers to the Self Test

1. (d)
2. (c)
3. (b)
4. (a)
5. (b)
6. (a)
7. (c)
8. (d)
9. (c)
10. (b)

Appendix

Appendix:

Specifications and Pinouts

Specifications and pinouts for the following components are given in this Appendix for your convenience. The TTL components can be used for the suggested experiments in this book. There are CMOS equivalents available.

6802 Microprocessor
6821 PIA
6850 ACIA
7400 Quad NAND Gates
7404 Hex Inverter
7408 Quad Two-Input AND Gate
7447 Seven-segment Decoder
7476 JK Flip Flops
7495 Register (PIPO)
74164 Register (SIPO)
74165 Register (PISO)
74181 ALU
74191 Programmable Counter

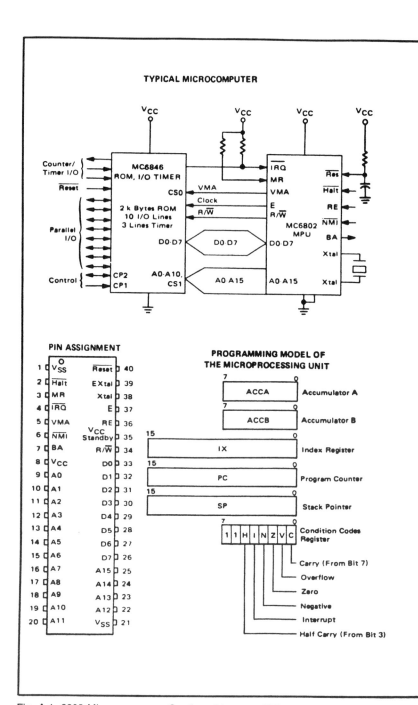

Fig. A-1. 6802 Microprocessor. Continued to page 187.

184

SAVING THE STATUS OF THE MICROPROCESSOR IN THE STACK

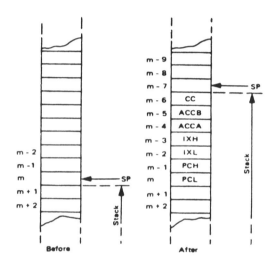

SP = Stack Pointer
CC = Condition Codes (Also called the Processor Status Byte)
ACCB = Accumulator B
ACCA = Accumulator A
IXH = Index Register, Higher Order 8 Bits
IXL = Index Register, Lower Order 8 Bits
PCH = Program Counter, Higher Order 8 Bits
PCL = Program Counter, Lower Order 8 Bits

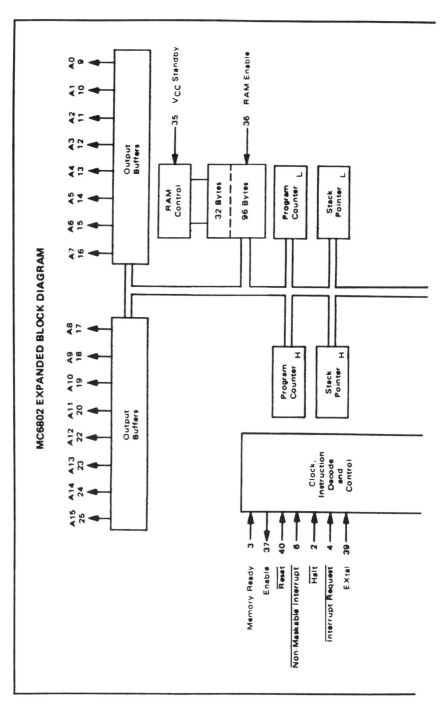

MC6802 EXPANDED BLOCK DIAGRAM

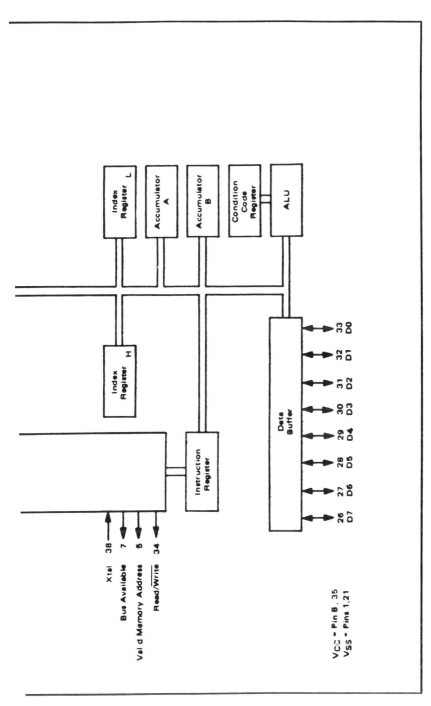

EXPANDED BLOCK DIAGRAM

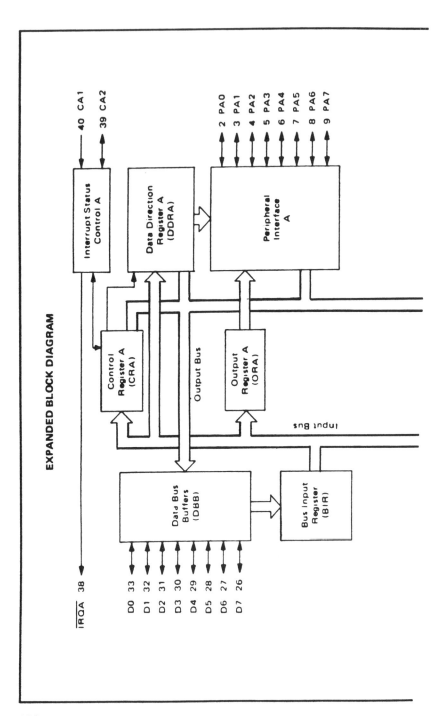

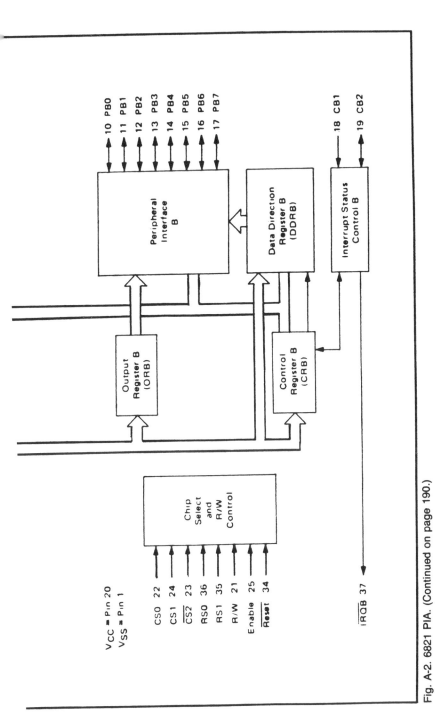

Fig. A-2. 6821 PIA. (Continued on page 190.)

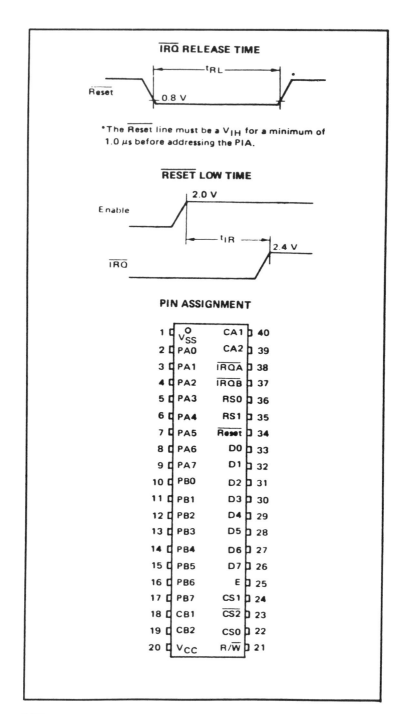

$\overline{\text{IRQ}}$ RELEASE TIME

Reset

t_{RL}

0.8 V

*The Reset line must be a V_{IH} for a minimum of 1.0 μs before addressing the PIA.

$\overline{\text{RESET}}$ LOW TIME

Enable

2.0 V

t_{IR}

2.4 V

$\overline{\text{IRQ}}$

PIN ASSIGNMENT

1	V_{SS}	CA1	40
2	PA0	CA2	39
3	PA1	$\overline{\text{IRQA}}$	38
4	PA2	$\overline{\text{IRQB}}$	37
5	PA3	RS0	36
6	PA4	RS1	35
7	PA5	Reset	34
8	PA6	D0	33
9	PA7	D1	32
10	PB0	D2	31
11	PB1	D3	30
12	PB2	D4	29
13	PB3	D5	28
14	PB4	D6	27
15	PB5	D7	26
16	PB6	E	25
17	PB7	CS1	24
18	CB1	$\overline{\text{CS2}}$	23
19	CB2	CS0	22
20	V_{CC}	R/$\overline{\text{W}}$	21

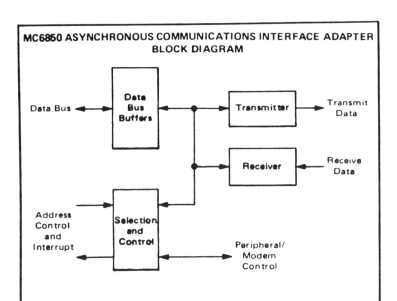

MC6850 ASYNCHRONOUS COMMUNICATIONS INTERFACE ADAPTER BLOCK DIAGRAM

DEFINITION OF ACIA REGISTER CONTENTS

Data Bus Line Number	Buffer Address			
	RS • R/W̄ Transmit Data Register	RS • R/W Receive Data Register	R̄S̄ • R/W̄ Control Register	R̄S̄ • R/W Status Register
	(Write Only)	(Read Only)	(Write Only)	(Read Only)
0	Data Bit 0*	Data Bit 0	Counter Divide Select 1 (CR0)	Receive Data Register Full (RDRF)
1	Data Bit 1	Data Bit 1	Counter Divide Select 2 (CR1)	Transmit Data Register Empty (TDRE)
2	Data Bit 2	Data Bit 2	Word Select 1 (CR2)	Data Carrier Detect (D̄C̄D̄)
3	Data Bit 3	Data Bit 3	Word Select 2 (CR3)	Clear to Send (C̄T̄S̄)
4	Data Bit 4	Data Bit 4	Word Select 3 (CR4)	Framing Error (FE)
5	Data Bit 5	Data Bit 5	Transmit Control 1 (CR5)	Receiver Overrun (OVRN)
6	Data Bit 6	Data Bit 6	Transmit Control 2 (CR6)	Parity Error (PE)
7	Data Bit 7***	Data Bit 7**	Receive Interrupt Enable (CR7)	Interrupt Request (IRQ)

* Leading bit = LSB : Bit 0
** Data bit will be zero in 7 bit plus parity modes
*** Data bit is "don't care" in 7 bit plus parity modes

Fig. A-3. 6850 ACIA. Continued to page 193.

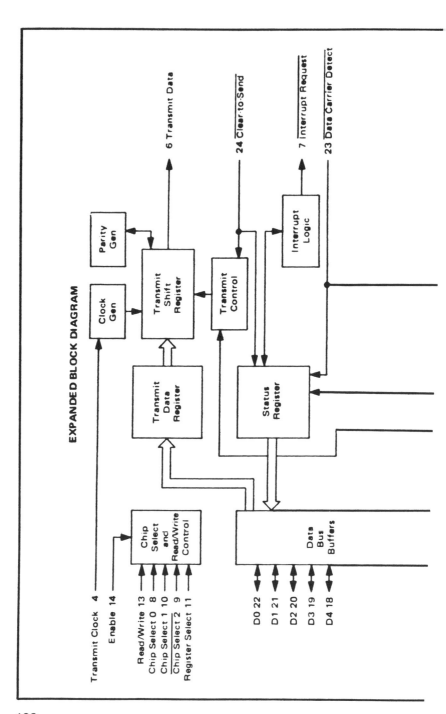

EXPANDED BLOCK DIAGRAM

192

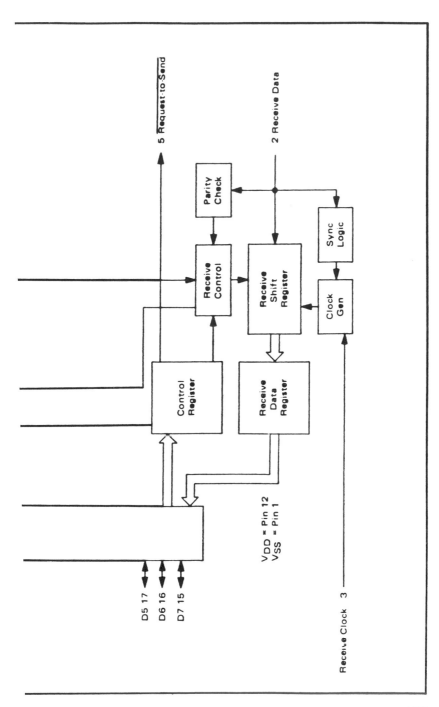

193

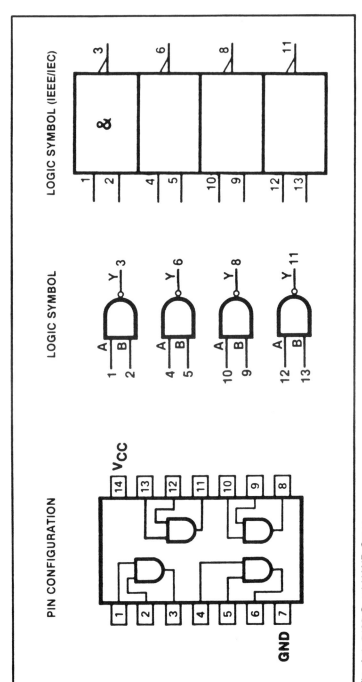

PIN CONFIGURATION

LOGIC SYMBOL

LOGIC SYMBOL (IEEE/IEC)

Fig. A-4. 7400 Quad NAND Gates.

194

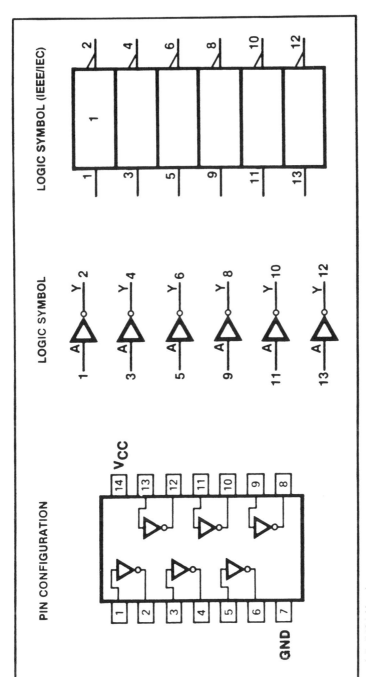

Fig. A.-5. 7404 Hex Inverter.

195

196

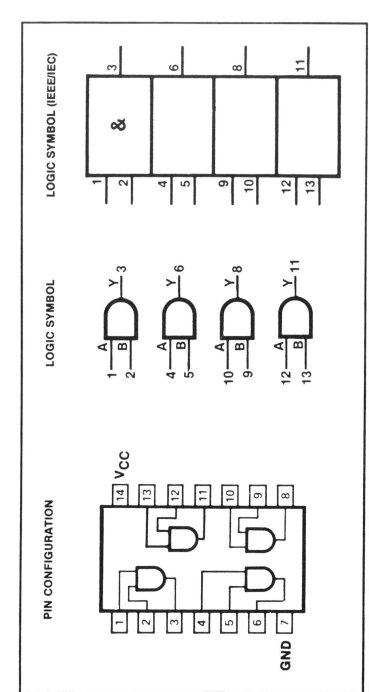

Fig. A-6. 7408 Quad Two-Input AND Gate.

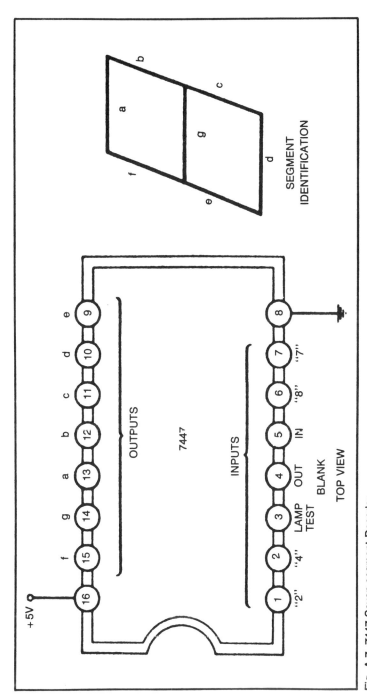

Fig. A-7. 7447 Seven-segment Decoder.

197

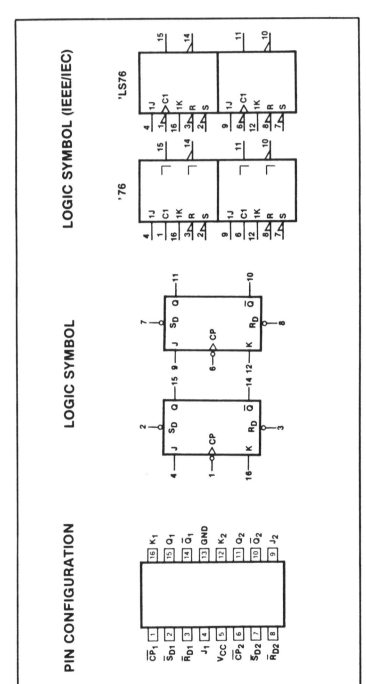

Fig. A-8. 7476 JK - Flip Flops.

198

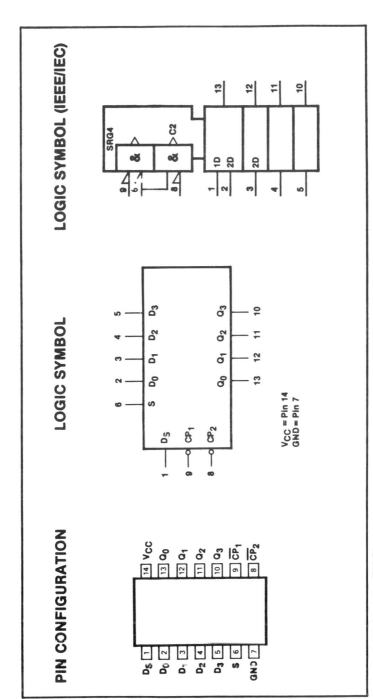

Fig. A-9. 7495 Register (PIPO).

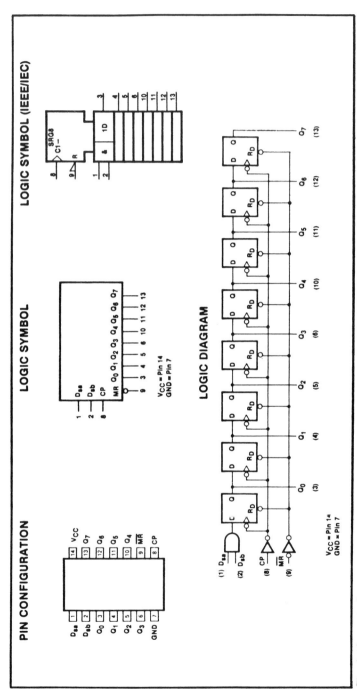

Fig. A-10. 74164 Register (SIPO).

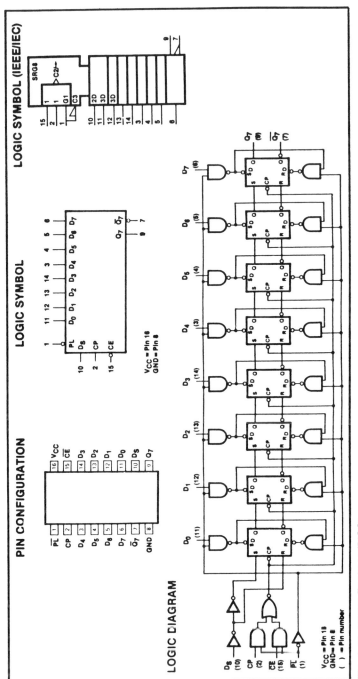

Fig. A-11. 74165 Register (PISO).

201

Mode select inputs				Active high inputs & outputs	
S3	S2		S0	Logic (M = H)	Arithmetic** (M = L) (C$_n$ = H)
L	L	L	L	$\overline{A}$	A
L	L	L	H	$\overline{A + B}$	A + B
L	L	H	L	$\overline{A}B$	A + $\overline{B}$
L	L	H	H	Logical 0	minus 1
L	H	L	L	$\overline{AB}$	A plus A$\overline{B}$
L	H	L	H	$\overline{B}$	(A + B) plus A$\overline{B}$
L	H	H	L	A ⊕ B	A minus B minus 1
L	H	H	H	A$\overline{B}$	AB minus 1
H	L	L	L	$\overline{A}$ + B	A plus AB
H	L	L	H	$\overline{A ⊕ B}$	A plus B
H	L	H	L	B	(A + $\overline{B}$) plus AB
H	L	H	H	AB	AB minus 1
H	H	L	L	Logical 1	A plus A*
H	H	L	H	A + $\overline{B}$	(A + B) plus A
H	H	H	L	A + B	(A + $\overline{B}$) plus A
H	H	H	H	A	A minus 1
L	L	L	L	$\overline{A}$	A minus 1
L	L	L	H	$\overline{AB}$	AB minus 1
L	L	H	L	$\overline{A}$ + B	A$\overline{B}$ minus 1
L	L	H	H	Logical 1	minus 1
L	H	L	L	$\overline{A + B}$	A plus (A + $\overline{B}$)
L	H	L	H	$\overline{B}$	AB plus (A + $\overline{B}$)
L	H	H	L	$\overline{A ⊕ B}$	A minus B minus 1
L	H	H	H	A + $\overline{B}$	A + B
H	L	L	L	$\overline{A}B$	A plus (A + B)
H	L	L	H	A ⊕ B	A plus B
H	L	H	L	B	A$\overline{B}$ plus (A + B)
H	L	H	H	A + B	A + B
H	H	L	L	Logical 0	A plus A*
H	H	L	H	A$\overline{B}$	AB plus A
H	H	H	L	AB	A$\overline{B}$ plus A
H	H	H	H	A	A

L = LOW voltage
H = HIGH voltage level
*Each bit is shifted to the next more significant position.
**Arithmetic operations expressed in 2s complement notation.

Fig. A-12. 74181 ALU.

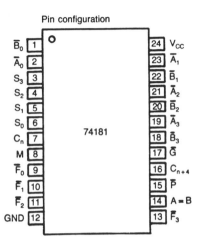

Pin configuration

74181

$\overline{B}_0$ 1 — 24 V_{CC}
$\overline{A}_0$ 2 — 23 $\overline{A}_1$
S_3 3 — 22 $\overline{B}_1$
S_2 4 — 21 $\overline{A}_2$
S_1 5 — 20 $\overline{B}_2$
S_0 6 — 19 $\overline{A}_3$
C_n 7 — 18 $\overline{B}_3$
M 8 — 17 $\overline{G}$
$\overline{F}_0$ 9 — 16 C_{n+4}
$\overline{F}_1$ 10 — 15 $\overline{P}$
$\overline{F}_2$ 11 — 14 A = B
GND 12 — 13 $\overline{F}_3$

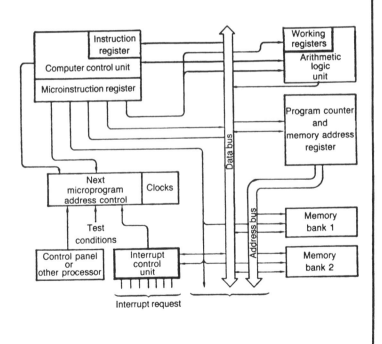

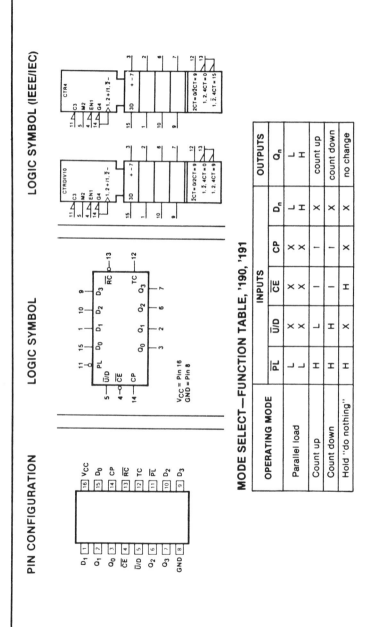

PIN CONFIGURATION

LOGIC SYMBOL

LOGIC SYMBOL (IEEE/IEC)

MODE SELECT—FUNCTION TABLE, '190, '191

OPERATING MODE	INPUTS					OUTPUTS
	$\overline{PL}$	$\overline{U}/D$	$\overline{CE}$	CP	D_n	Q_n
Parallel load	L	X	X	X	L	L
	L	X	X	X	H	H
Count up	H	L	L	↑	X	count up
Count down	H	H	L	↑	X	count down
Hold "do nothing"	H	X	H	X	X	no change

204

TC AND $\overline{RC}$ FUNCTION TABLE, '190

INPUTS			TERMINAL COUNT STATE				OUTPUTS	
$\overline{U}/D$	$\overline{CE}$	CP	Q_0	Q_1	Q_2	Q_3	TC	$\overline{RC}$
H	H	X	H	X	X	H	L	H
L	H	X	H	X	X	H	H	H
L	L	⊔	H	X	X	H	⌐	⊔
L	H	X	L	L	L	L	L	H
H	H	X	L	L	L	L	H	H
H	L	⊔	L	L	L	L	⌐	⊔

TC AND $\overline{RC}$ FUNCTION TABLE, '191

INPUTS			TERMINAL COUNT STATE				OUTPUTS	
$\overline{U}/D$	$\overline{CE}$	CP	Q_0	Q_1	Q_2	Q_3	TC	$\overline{RC}$
H	H	X	H	H	H	H	L	H
L	H	X	H	H	H	H	H	H
L	L	⊔	H	H	H	H	⌐	⊔
L	H	X	L	L	L	L	L	H
H	H	X	L	L	L	L	H	H
H	L	⊔	L	L	L	L	⌐	⊔

H = HIGH voltage level steady state.
L = LOW voltage level steady state.
l = LOW voltage level one setup time prior to the LOW-to-HIGH clock transition.
X = Don't care.
↑ = LOW-to-HIGH clock transition.
⊔ = LOW pulse.
⌐ = TC goes LOW on a LOW-to-HIGH clock transition.

Fig. A-13. 74191 Programmable Counter.

Index

Index

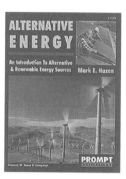

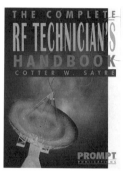

Alternative Energy
Mark E. Hazen

The Complete RF Technician's Handbook
Cotter W. Sayre

This book is designed to introduce readers to the many different forms of energy mankind has learned to put to use. Generally, energy sources are harnessed for the purpose of producing electricity. This process relies on transducers to transform energy from one form into another. *Alternative Energy* will not only address transducers and the five most common sources of energy that can be converted to electricity, it will also explore solar energy, the harnessing of the wind for energy, geothermal energy, and nuclear energy.

The *Complete RF Technician's Handbook* will furnish the working technician or student with a solid grounding in the latest methods and circuits employed in today's RF communications gear. It will also give readers the ability to test and troubleshoot transmitters, transceivers, and receivers with absolute confidence. Some of the topics covered include reactance, phase angle, logarithms, diodes, passive filters, amplifiers, and distortion. Various multiplexing methods and data, satellite, spread spectrum, cellular, and microwave communication technologies are discussed.

Professional Reference
320 pages ✦ Paperback ✦
7-3/8 x 9-1/4"
ISBN: 0-7906-1079-5 ✦ Sams:
61079
$18.95 ($25.95 Canada) ✦ October 1996

Professional Reference
281 pages ✦ Paperback ✦
8-1/2 x 11"
ISBN: 0-7906-1085-X ✦ Sams:
61085
$24.95 ($33.95 Canada) ✦ July 1996

PC Hardware Projects
Volume 1
James "J.J." Barbarello

PC Hardware Projects
Volume 2
James "J.J." Barbarello

Now you can create your own PC-based digital design workstation! Using commonly available components and standard construction techniques, you can build some key tools to troubleshoot digital circuits and test your printer, fax, modem, and other multi-conductor cables.

This book will guide you through the construction of a channel logic analyzer, and a multipath continuity tester. You will also be able to combine the projects with an appropriate power supply and a prototyping solderless breadboard system into a single digital workstation interface! **PROJECT SOFTWARE DISK INCLUDED!**

PC Hardware Projects, Volume 2, discusses stepper motors and how to control them. It investigates different methods to control stepper motors, and provides you with circuitry for a dedicated IC controller and a discrete component hardware controller. Then, this book guides you through every step of constructing an automated, PC-controlled drilling machine. You'll then walk through an actual design layout, creating a PC design and board. With the help of the information and the data file disk included, you'll have transformed your PC into your very won PCB fabrication house! **PROJECT SOFTWARE DISK INCLUDED!**

Computer Technology
256 pages ◆ Paperback ◆
7-3/8 x 9-1/4"
ISBN: 0-7906-1104-X ◆ Sams:
61104
$24.95 ◆ Feb. 1997

Computer Technology
256 pages + Paperback +
7-3/8 x 9-1/4"
ISBN: 0-7906-1109-0 ◆ Sams:
61109
$24.95 ◆ May 1997

Semiconductor Cross Reference Book Fourth Edition

Howard W. Sams & Company

This newly revised and updated reference book is the most comprehensive guide to replacement data available for engineers, technicians, and those who work with semiconductors. With more than 490,000 part numbers, type numbers, and other identifying numbers listed, technicians will have no problem locating the replacement or substitution information needed. There is not another book on the market that can rival the breadth and reliability of information available in the fourth edition of the *Semiconductor Cross Reference Book.*

Professional Reference
688 pages ✦ Paperback ✦
8-1/2 x 11"
ISBN: 0-7906-1080-9 ✦ Sams: 61080
$24.95 ($33.95 Canada) ✦ August 1996

IC Cross Reference Book Second Edition

Howard W. Sams & Company

The engineering staff of Howard W. Sams & Company assembled the *IC Cross Reference Book* to help readers find replacements or substitutions for more than 35,000 ICs and modules. It is an easy-to-use cross reference guide and includes part numbers for the United States, Europe, and the Far East. This reference book was compiled from manufacturers' data and from the analysis of consumer electronics devices for PHOTOFACT® service data, which has been relied upon since 1946 by service technicians worldwide.

Professional Reference
192 pages ✦ Paperback ✦
8-1/2 x 11"
ISBN: 0-7906-1096-5 ✦ Sams: 61096
$19.95 ($26.99 Canada) ✦
November 1996

**CALL 1-800-428-7267 TODAY FOR THE NAME OF
YOUR NEAREST PROMPT PUBLICATIONS DISTRIBUTOR**

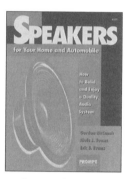

Speakers for Your Home & Automobile

Gordon McComb, Alvis J. Evans, & Eric J. Evans

The cleanest CD sound, the quietest turntable, or the clearest FM signal are useless without a fine speaker system. This book not only tells readers how to build quality speaker systems, it also shows them what components to choose and why. The comprehensive coverage includes speakers, finishing touches, construction techniques, wiring speakers, and automotive sound systems.

Audio Technology
164 pages ♦ Paperback ♦ 6 x 9"
ISBN: 0-7906-1025-6 ♦ Sams: 61025
$14.95 ($20.95 Canada) ♦ November 1992

Sound Systems for Your Automobile

Alvis J. Evans & Eric J. Evans

This book provides the average vehicle owner with the information and skills needed to install, upgrade, and design automotive sound systems. From terms and definitions straight up to performance objectives and cutting layouts, *Sound Systems* will show the reader how to build automotive sound systems that provide occupants with live performance reproductions that rival home audio systems.

Whether starting from scratch or upgrading, this book uses easy-to-follow steps to help readers plan their system, choose components and speakers, and install and interconnect them to achieve the best sound quality possible.

Audio Technology
124 pages ♦ Paperback ♦ 6 x 9"
ISBN: 0-7906-1046-9 ♦ Sams: 61046
$16.95 ($22.99 Canada) ♦ January 1994

**CALL 1-800-428-7267 TODAY FOR THE NAME OF
YOUR NEAREST PROMPT PUBLICATIONS DISTRIBUTOR**

Is This Thing On?

Gordon McComb

Theory & Design of Loudspeaker Enclosures

Dr. J. Ernest Benson

Is This Thing On? takes readers through each step of selecting components, installing, adjusting, and maintaining a sound system for small meeting rooms, churches, lecture halls, public-address systems for schools or offices, or any other large room.

In easy-to-understand terms, drawings and illustrations, *Is This Thing On?* explains the exact procedures behind connections and troubleshooting diagnostics. With the help of this book, hobbyists and technicians can avoid problems that often occur while setting up sound systems for events and lectures.

The design of loudspeaker enclosures, particularly vented enclosures, has been a subject of continuing interest since 1930. Since that time, a wide range of interests surrounding loudspeaker enclosures have sprung up that grapple with the various aspects of the subject, especially design. *Theory & Design of Loudspeaker Enclosures* lays the groundwork for readers who want to understand the general functions of loudspeaker enclosure systems and eventually experiment with their own design.

Audio Technology
136 pages ♦ Paperback ♦ 6 x 9"
ISBN: 0-7906-1081-7 ♦ Sams:
61081
$14.95 ($20.95 Canada) ♦ April
1996

Audio Technology
244 pages ♦ Paperback ♦ 6 x 9"
ISBN: 0-7906-1093 0 ♦ Sams:
61093
$19.95 ($26.99 Canada) ♦ August
1996

**CALL 1-800-428-7267 TODAY FOR THE NAME OF
YOUR NEAREST PROMPT PUBLICATIONS DISTRIBUTOR**

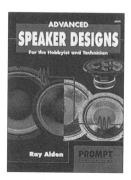

Advanced Speaker Designs
Ray Alden

Making Sense of Sound
Alvis J. Evans

Advanced Speaker Designs shows the hobbyist and the experienced technician how to create high-quality speaker systems for the home, office, or auditorium. Every part of the system is covered in detail, from the driver and crossover network to the enclosure itself. Readers can build speaker systems from the parts lists and instructions provided, or they can actually learn to calculate design parameters, system responses, and component values with scientific calculators or PC software.

This book deals with the subject of sound — how it is detected and processed using electronics in equipment that spans the full spectrum of consumer electronics. It concentrates on explaining basic concepts and fundamentals to provide easy-to-understand information, yet it contains enough detail to be of high interest to the serious practitioner. Discussion begins with how sound propagates and common sound characteristics, before moving on to the more advanced concepts of amplification and distortion. *Making Sense of Sound* was designed to cover a broad scope, yet in enough detail to be a useful reference for readers at every level.

Audio Technology
136 pages ✦ Paperback ✦ 6 x 9"
ISBN: 0-7906-1070-1 ✦ Sams: 61070
$16.95 ($22.99 Canada) ✦ July 1995

Audio Technology
112 pages ✦ Paperback ✦ 6 x 9"
ISBN: 0-7906-1026-4 ✦ Sams: 61026
$10.95 ($14.95 Canada) ✦ November 1992

CALL 1-800-428-7267 TODAY FOR THE NAME OF YOUR NEAREST PROMPT PUBLICATIONS DISTRIBUTOR

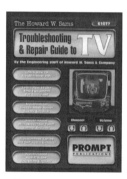

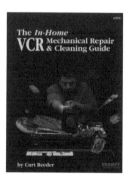

The Howard W. Sams Troubleshooting & Repair Guide to TV

Howard W. Sams & Company

The Howard W. Sams Troubleshooting & Repair Guide to TV is the most complete and up-to-date television repair book available. Included in its more than 300 pages is complete repair information for all makes of TVs, time-saving features that even the pros don't know, comprehensive basic electronics information, and extensive coverage of common TV symptoms.

This repair guide is completely illustrated with useful photos, schematics, graphs, and flowcharts. It covers audio, video, technician safety, test equipment, power supplies, picture-in-picture, and much more.

Video Technology
384 pages Paperback ♦
8-1/2 x 11"
ISBN: 0-7906-1077-9 ♦ Sams:
61077
$29.95 ($39.95 Canada) ♦ June
1996

The In-Home VCR Mechanical Repair & Cleaning Guide

Curt Reeder

· Like any machine that is used in the home or office, a VCR requires minimal service to keep it functioning well and for a long time. However, a technical or electrical engineering degree is not required to begin regular maintenance on a VCR. *The In-Home VCR Mechanical Repair & Cleaning Guide* shows readers the tricks and secrets of VCR maintenance using just a few small hand tools, such as tweezers and a power screwdriver.

This book is also geared toward entrepreneurs who may consider starting a new VCR service business of their own.

Video Technology
222 pages ♦ Paperback ♦
8-3/8 x 10-7/8"
ISBN: 0-7906-1076-0 ♦ Sams:
61076
$19.95 ($26.99 Canada) ♦ April
1996

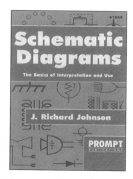

Dear Reader: *We'd like your views on the books we publish.*

PROMPT® Publications, a division of Howard W. Sams & Company (A Bell Atlantic Company), is dedicated to bringing you timely and authoritative documentation and information you can use. You can help us in our continuing effort to meet your information needs. Please take a few moments to answer the questions below. Your answers will help us serve you better in the future.

1. What is the title of the book you purchased?_____
2. Where do you usually buy books?_____
3. Where did you buy this book?_____
4. What did you like most about the book?_____
5. What did you like least?_____
6. Is there any other information you'd like included?_____
7. In what subject areas would you like us to publish more books? (Please check the boxes next to your fields of interest.)

❑ Audio Equipment Repair   ❑ Home Appliance Repair

❑ Camcorder Repair   ❑ Mobile Communications

❑ Computer Repair   ❑ Security Systems

❑ Electronic Concepts Theory   ❑ Sound System Installation

❑ Electronic Projects/Hobbies   ❑ TV Repair

❑ Electronic Reference   ❑ VCR Repair

8. Are there other subjects that you'd like to see books about? _____

9. Comments _____
_____
_____

• •

Name _____
Address _____
City _____ State/ZIP _____
E-Mail _____

Would you like a *FREE* PROMPT® Publications catalog?   ❑Yes   ❑ No

Thank you for helping us make our books better for all of our readers. Please drop this postage-paid card into the nearest mailbox.

For more information about PROMPT® Publications, see your authorized Howard Sams distributor or call 1-800-428-7267 for the name of your nearest PROMPT® Publications distributor.

PROMPT.
PUBLICATIONS

A Division of *Howard W. Sams & Company*
A Bell Atlantic Company
2647 Waterfront Parkway, East Dr.
Indianapolis, IN 46214-2041